AXXX$6.50

Book 1

Germ Cells and Fertilization

REPRODUCTION IN MAMMALS

Book 1

Germ Cells and Fertilization

EDITED BY

C. R. AUSTIN

Fellow of Fitzwilliam College,
Charles Darwin Professor of Animal Embryology,
University of Cambridge

AND

R. V. SHORT

Fellow of Magdalene College,
Reader in Reproductive Biology,
University of Cambridge

ILLUSTRATIONS BY JOHN R. FULLER

CAMBRIDGE UNIVERSITY PRESS

Cambridge

London New York Melbourne

Published by the Syndics of the Cambridge University Press
The Pitt Building, Trumpington Street, Cambridge CB2 1RP
Bentley House, 200 Euston Road, London NW1 2DB
32 East 57th Street, New York, NY 10022, USA
296 Beaconsfield Parade, Middle Park, Melbourne 3206, Australia

© Cambridge University Press 1972

Library of Congress catalogue card number: 73-174261

ISBN 0 521 08408 3 hard covers
ISBN 0 521 09690 1 paperback

First published 1972
Reprinted 1973, 1978

First printed by offset in Great Britain by
Alden & Mowbray Ltd
at the Alden Press, Oxford
Reprinted in Malta by Interprint (Malta) Ltd

Contents

Contributors to Book 1

C. R. Austin
Physiological Laboratory
Downing Street
Cambridge

T. G. Baker
Hormone Laboratory
Department of Obstetrics and Gynaecology
University of Edinburgh

V. Monesi
Universita di Roma
Istituto di Istologia ed Embriologia Generale
Viale Regina Elena, 289
Rome, Italy

R. M. F. S. Sadleir
Department of Biological Sciences
Simon Fraser University
British Columbia, Canada

Preface

Reproduction in Mammals is intended to meet the needs of undergraduates reading Zoology, Biology, Physiology, Medicine, Veterinary Science and Agriculture, and as a source of information for advanced students and research workers. It is published as a series of five small textbooks dealing with all major aspects of mammalian reproduction. Each of the component books is designed to cover independently fairly distinct subdivisions of the subject, so that readers can select texts relevant to their particular interests and needs, if reluctant to purchase the whole work. The contents lists of all the books are set out on the next page.

The first book is concerned with *Germ Cells and Fertilization* and with the reproductive cycles in which these are involved. The story begins with the primordial germ cells and their early history, and continues with the events that lead to the formation and union of gametes, and the establishment of the zygote. Some of the environmental influences that affect these processes are discussed in the fourth chapter, but the main controlling mechanisms are set out in later books.

Books in this series

Book 1. Germ Cells and Fertilization
Primordial germ cells. T. G. Baker
Oogenesis and ovulation. T. G. Baker
Spermatogenesis and the spermatozoa. V. Monesi
Cycles and seasons. R. M. F. S. Sadleir
Fertilization. C. R. Austin

Book 2. Embryonic and Fetal Development
The embryo. Anne McLaren
Sex determination and differentiation. R. V. Short
The fetus and birth. G. C. Liggins
Manipulation of development. R. L. Gardner
Pregnancy losses and birth defects. C. R. Austin

Book 3. Hormones in Reproduction
Reproductive hormones. D. T. Baird
The hypothalamus. B. A. Cross
Role of hormones in sex cycles. R. V. Short
Role of hormones in pregnancy. R. B. Heap
Lactation and its hormonal control. Alfred T. Cowie

Book 4. Reproductive Patterns
Species differences. R. V. Short
Behavioural patterns. J. Herbert
Environmental effects. R. M. F. S. Sadleir
Immunological influences. R. G. Edwards
Ageing and reproduction. C. E. Adams

Book 5. Artificial Control of Reproduction
Increasing reproductive potential in farm animals. C. Polge
Limiting human reproductive potential. D. M. Potts
Chemical methods of male contraception. Harold Jackson
Control of human development. R. G. Edwards
Reproduction and human society. R. V. Short
The ethics of manipulating reproduction in man. C. R. Austin

1 Primordial germ cells

T. G. Baker

One of the most important concepts in reproductive physiology is that the definitive germ cells – the eggs and spermatozoa – are derived solely from the primitive sex cells found early in embryonic development. We can express this idea in another way, namely, that there is a continuity of the germ-cell line from embryo to adult. Perhaps the best evidence in support of this concept stems from studies of certain frogs and toads where a localized region of the vegetal (yolky) pole in the fertilized egg can be identified as the 'germinal cytoplasm'. This material, which is rich in ribonucleic acid, can be traced through successive cleavage divisions of the egg so that by the blastula stage some 2–14 cells are found to contain 'germinal cytoplasm'. During the subsequent morphogenetic movements within the embryo (gastrulation) these primitive sex cells become situated deep in the endoderm, from whence they migrate to the developing gonads. If the primordial germ cells fail to reach the gonadal ridges, or if they are destroyed by surgical removal or by irradiation during migration, the resulting ovaries will be sterile since they will contain no sex cells.

Selective destruction of the primordial germ cells is difficult to carry out in mammalian embryos owing to the lack of 'germinal cytoplasm' and the fact that it is difficult to get at the fetus within the uterus. Nevertheless, histological studies of gonadal development in mice, rats, cattle and man have provided parts of the story from which the process as a whole can be deduced. Until recent years the results were often thought to be inconclusive and hence the origin of germ cells has been the subject of unceasing debate for some 50 years or more. Some authors disputed the germinal nature of the primitive sex cells altogether and suggested that the definitive germ cells were formed from somatic cells during each reproductive cycle. Others, while accepting the initial role of primordial germ cells in gonadal differentiation, believed that the early sex cells degenerated to be replaced by cells arising in the so-called 'germinal' (coelomic) epithelium covering the developing gonads. As we shall see later, recent studies provide overwhelming evidence in support of the classical view of the continuity of the germ cell line, without disclosing any evidence for the transformation of epithelial (somatic) cells into those of the germinal line.

ORIGIN AND MIGRATION OF THE SEX CELLS

Primordial germ cells are larger than somatic cells and possess large round nuclei containing prominent nucleoli (Fig. 1-1).

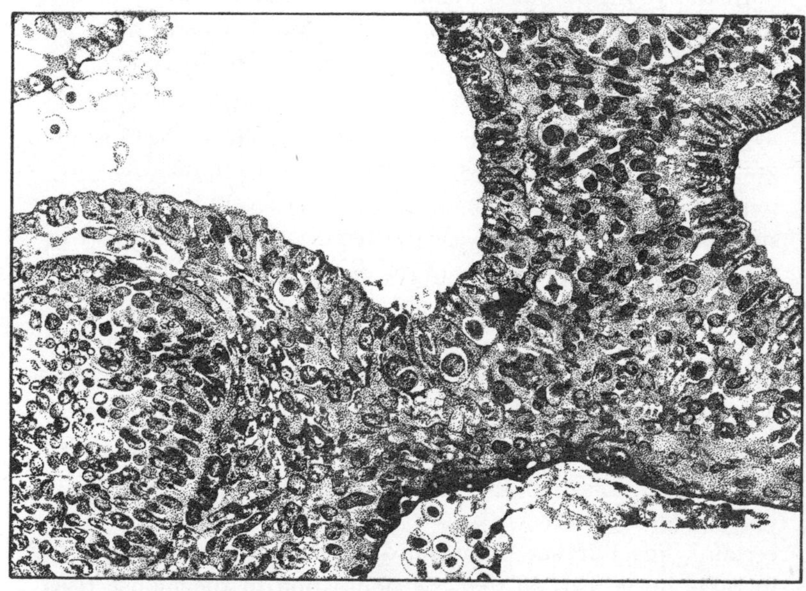

Fig. 1-1. Section of human embryo aged about 31 days showing primordial germ cells in the wall of the hind gut and in the gut mesentery. The cells are larger than the surrounding somatic cells and one (arrow) is at mitotic metaphase. (Drawn from E. Witschi, *Contr. Embryol.* **32**, 69, Fig. 10, Plate 3 (1948). By courtesy of the Carnegie Institution of Washington.)

Nevertheless, they can be confused with other cells in the body, especially when these are undergoing cell division (mitosis). The primitive sex cells are more precisely identified by certain histochemical techniques since they contain specific enzymes and chemical substances such as alkaline phosphatase, glycogen and esterases. In practice, embryologists use a variety of histological criteria when distinguishing primordial germ cells from other cells in the embryo.

The best accounts of the origin and migration of germ cells are those for man himself and so these will be described in some detail. One might add that essentially similar results have been obtained for embryos of the mouse, rat, rabbit, cow and Rhesus monkey, although the timing of the individual stages varies between species.

In the youngest human embryos available for study (about the 3rd week after conception), the primordial germ cells are situated in the epithelium of the yolk sac near to the developing allantois (see Fig. 1-2). By the time that the embryo has developed about 13–20 somites (segments) the cells have migrated to the connective tissue of the hind gut from whence they move into the gut mesentery (see Fig. 1-1). From the 25-somite stage onwards (about 30 days after fertilization), the majority of the cells pass into the region of the developing kidneys and then into the adjacent gonadal primordia (genital ridges).

The primordial germ cells possess pseudopodia which they use to wriggle through the tissues of the yolk sac and gut. Migration is thus an active process involving locomotion similar to that of an amoeba. The early phases of migration may also involve the digestion of membranes and other obstacles by lytic enzymes produced by the germ cells, but this is not yet certain. The number of primordial germ cells is believed to increase by mitosis so that the population in the mouse rises from about 100 early in migration to some 5000 in the fully differentiated gonad. Occasional dividing cells can be seen in histological sections of human embryos (see Fig. 1-1), and once the germ cells have completed their colonization of the gonad a dramatic increase in cell number occurs, especially in the female.

The genital ridges in the dorsal body wall of the embryo have been thought to produce a so-far unidentified chemotactic substance ('telopheron') which attracts the primitive sex cells. The only evidence for such a substance comes from studies of gonadal cells in tissue culture. Germ cells cultured on their own undergo random movements, whereas the somatic cells remain essentially stationary. However, if genital ridge tissue is placed at the edge of the culture dish the germ cells are 'attracted' towards it. And even if gonadal tissue is transplanted to ab-

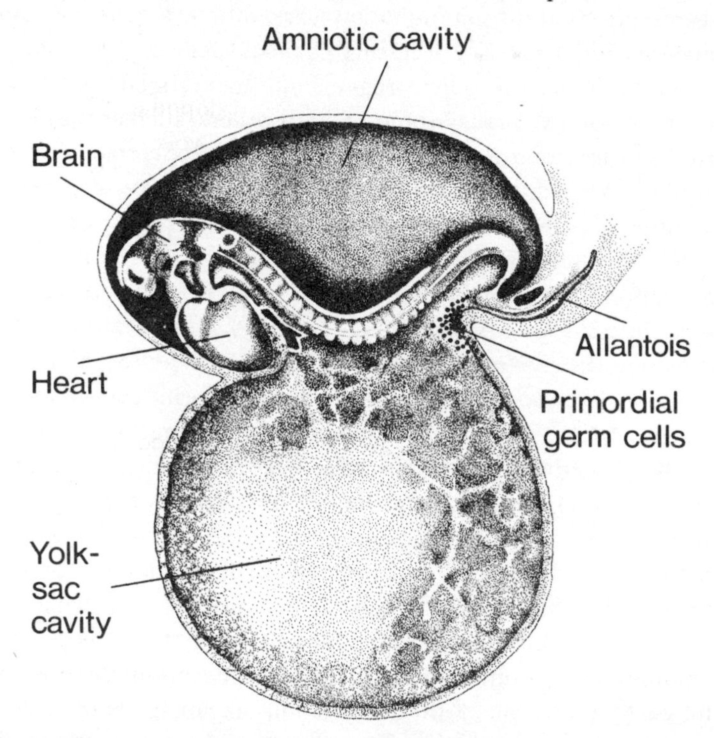

Fig. 1-2. Reconstruction of 24-day human embryo in its amnion. The primordial germ cells (black dots) are grouped at the top of the yolk sac and in the ventral wall of the developing hind gut. (Re-drawn after E. Witschi, *Contr. Embryol.* **32**, 69, Fig. 1, Plate 1 (1948). By courtesy of the Carnegie Institution of Washington.)

normal sites within the embryo, the germ cells can still migrate into it quite normally. The two genital ridges may produce different amounts of telopheron, since more germ cells reach the left gonads of birds than the right. Discrepancies in the numbers of germ cells in the right and left gonads of mammals are also well documented although the variation is less evident than in the chick. If such a chemotactic substance as telopheron exists it is not species specific, since the developing gonads of the chick can attract both chick and mouse primordial germ cells.

Not all the sex cells reach the gonadal ridges; some are seemingly 'lost' during migration while others degenerate along the way. Small areas of accessory gonadal tissue are occasionally found deep in the pelvic region and in vestigial structures. Furthermore, in some chromosomal disorders, such as the loss of an X-chromosome in women and also the case of certain mutants of mice, the primordial germ cells or their derivatives degenerate after completing migration.

We have seen previously that the normal process of germ-cell migration in mammals involves amoeboid activity and possibly also chemotactic substances and lytic enzymes. But the results of recent studies indicate that at least some of the cells may occasionally travel passively via the blood stream and 'home-in' on the gonad. This type of migration is common in birds and in the primitive reptile *Sphenodon punctatus* and may occur in certain mammals (cow, pig, sheep, goat: see Book 2, Chapter 2). It would also allow an exchange of germ cells between dizygotic (non-identical) twins in which placental circulations become fused before the germ cells reach the gonadal primordia. We know that fusion may lead to the passage of blood cells from one individual to the other, the embryos becoming 'chimaeras' for each other's blood groups. In addition, some of the primordial germ cells in developing calf twins may pass across the placenta in the blood so that these twins are also germ-cell chimaeras. These points are discussed more fully in Book 2, especially Chapter 2: the general consensus of opinion is that the passage of germ cells via the blood stream in mammals is an accidental rather than a regular occurrence.

COLONIZATION OF THE PRESUMPTIVE GONAD

The presumptive gonads (gonadal primordia, ridges or *anlagen*) each consist of a layer of thickened coelomic epithelium overlying a 'knot' of mesenchymal tissue (embryonic mesoderm). Towards the end of migration the germ cells move from the developing kidneys (mesonephros) into the genital ridges and take up either a cortical or medullary position. In general morphological terms, the germ cells remain in the cortex if the gonad is to differentiate into an ovary, while the passage of such cells into the medulla (mesenchyme) is associated with the development of a testis. However, this pattern of gonadal differentiation can be modified by factors within the embryo such as hormones and non-hormonal inductor substances. Some germ cells in the male gonad remain in the cortex, while a few in the female may reach the medulla (e.g. at the mesenteric border or hilum of the ovary). The germ cells that find themselves in the 'wrong' position in the gonad persist for long periods before undergoing degeneration; in some cases (e.g. in the striped skunk) they may even undergo division or meiotic changes, particularly in the female where treatment with gonadotrophic hormones is said to induce follicular growth and ovulation. These events seem to be rare and most of the misplaced cells, especially those in intersexes and interspecific hybrids, are eliminated shortly after birth.

In the early stages of gonadal differentiation the sex of the embryo can only be determined from cytogenetic studies; in terms of gross morphology and histology the gonad is 'indifferent' and bipotential since its differentiation can lead to the formation of either an ovary or a testis (see Book 2, Chapter 2).

THE MORPHOLOGY OF SEX DIFFERENTIATION

The indifferent gonad contains all the cellular components

necessary for its differentiation into either testis or ovary, namely, (1) medullary tissue, which develops into the medullary cords of the testis, (2) cortical tissue, formed from the coelomic epithelium, which may develop as the secondary sex cords to provide the substance of the ovary, (3) mesenchyme, which contributes the rest of the gonad in both sexes (including thecal and interstitial tissue); and (4) the primordial germ cells which have an extra-gonadal origin and develop into oogonia or spermatogonia according to their genetic constitution. The primitive kidney (mesonephros) may also contribute cells to the developing gonad and may be essential for its sexual differentiation; *in vitro* the latter process only occurs when the gonad is cultured together with the mesonephric blastema.

The process of sex differentiation largely consists in the development of a testis, or the lack of such changes. Thus, if the individual develops as a male, the germ cells pass into the medulla which develops at the expense of the cortex. The primary sex cords enclose the germ cells to become seminiferous tubules, and the formation of a connective tissue membrane (tunica albuginea) beneath the coelomic epithelium suppresses cortical development.

By contrast, ovarian development occurs much later and the early female gonad is recognized largely by the fact that it does not resemble a testis. Differentiation of the ovary is labile since transplantation into hosts of either sex results at the very least in the partial development of its medulla along the male line, if not in the production of an ovotestis. The development of the mammalian testis is largely determinate, however, and is unaffected by transplantation into developing females.

It would be wrong to assume from the foregoing paragraphs that the germ cells play no part in the differentiation of the gonad. The problem is a complex one – it is discussed in Chapter 2.

Certainly the position of the germ cells within the gonad affects development, female germ cells that enter the medulla usually degenerating or remaining dormant: there is no evidence in mammals that they ever become male germ cells. Similarly, male gonocytes that persist in the cortex are unlikely ever to become oocytes. This is in marked contrast to the situation in lower vertebrates where germ cells become spermatocytes or oocytes according to the genetic sex of the gonads in which they reside; sex reversal is easily accomplished by the administration of steroids or by such environmental factors as changes in temperature. It will be shown in Book 2 that environmental factors do not affect gonadal differentiation in mammals, and sex reversal has only been achieved in the Virginian opossum. It is possible that the development of mammalian gonads requires an adequate supply of fetal hormones at certain critical stages of differentiation but maternal hormones and exogenous steroids have little or no effect.

TRANSFORMATION INTO DEFINITIVE GERM CELLS

Towards the end of sex differentiation the number of primordial germ cells increases sharply by mitosis. The testes of rats have been shown to contain more sex cells at this early stage than do ovaries. The situation rapidly reverses, however, since the male germ cells become relatively undifferentiated gonocytes which rarely undergo division. The first spermatogonia appear in the human testis during the 5th month of intra-uterine life and in other species sometime after birth. The appearance of the testis, other than for an increase in size, changes only slightly before puberty and the spermatogenic tubules have little or no lumen (see Fig. 1-3).

By contrast, the ovary rapidly takes on the form of the adult gland, albeit diminutive (shortly after birth in most species; see Fig. 1-4). The ovarian germ cells rapidly become transformed into oogonia, which are larger and have a different distribution of cytoplasmic organelles to the primordial germ cells. The population of oogonia increases greatly in a relatively short period of time. Thus in the rat the number of germ cells increases from 11 500 to 75 500 during the period of oogonial

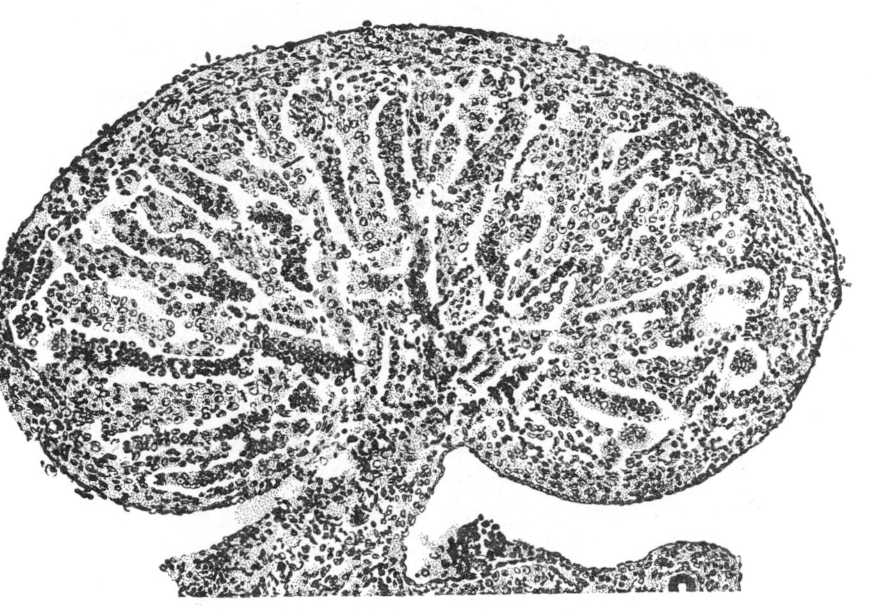

Fig. 1-3. Appearance of the testis at the end of the period of sex differentiation. The spermatic tubules have little or no lumen and few germ cells can be detected. (From G. von Wagenen and M. E. Simpson, *Embryology of the Ovary and Testis*, Plate 53, Fig. C. Yale University Press (1965).)

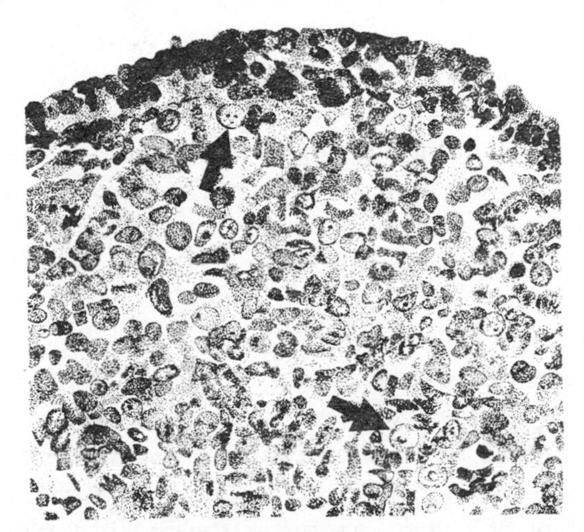

Fig. 1-4. Ovary at a similar stage to Fig. 1-3. The cortex is well developed and contains germ cells (two arrowed) at different stages of mitosis. (Source as Fig. 1-3, Plate 53, Fig. C.)

division (by comparison there are in man about 1700 cells during migration, 600 000 during the 2nd month of pregnancy, and almost 7 million at the 5th month; see Fig. 1-5). After what appears to be a finite number of mitoses, the oogonia become transformed into oocytes, when they enter upon the prophase of the first of two meiotic divisions (see Chapter 2). From this

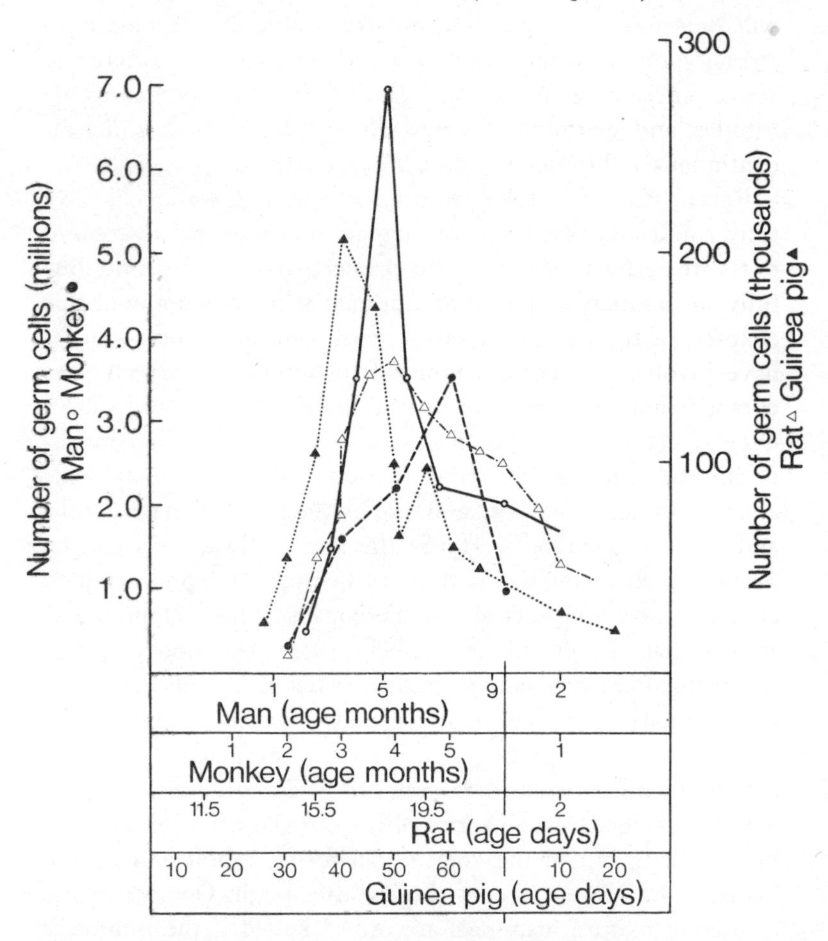

Fig. 1-5. Fluctuations in the number of germ cells in the ovaries of embryonic and fetal rats, guinea pigs, monkeys and human beings. (From T. G. Baker. Ch. 10 in *Reproductive Biology*. Ed. H. Balin and S. R. Glasser. Excerpta Medica (1971).)

time onwards all the oocytes are incapable of increasing their numbers and hence, when the oogonia are finally eliminated from the ovary (before birth in most species; see page 10), the population of germ cells can only be *reduced* with increasing age, and this occurs by the processes of atresia and ovulation (Chapter 2). Agents that damage or destroy the oocytes (e.g. radiation) will thus have a permanent and irrevocable effect. This is in strong contrast to the situation in the testis where mitotically active spermatogonia persist basally in the spermatogenic tubules, and spermatocytes pass through the phases of meiosis continuously throughout adult life (see Chapter 3).

Recent studies on the replication of DNA in germ cells have fully confirmed that the definitive germ cells in the ovary and testis are derived solely from the primordial germ cells, and thus fully substantiate the concept, outlined at the beginning of this chapter, of the continuity of the germ-cell line. These studies have involved injecting a radioactive isotope of a DNA precursor (tritiated thymidine) into laboratory animals and subsequently detecting the site of uptake of the substance by means of autoradiography. The label is taken up by all somatic cells about to undergo division, and also by the primordial germ cells and gonadal stem cells. Hence the fate of these cells can be proved by observing the pattern of labelling in the photographic emulsion overlying sections of the gonads. This technique has shown that the gonadal stem cells (oogonia, gonocytes and spermatogonia) arise only from primordial germ cells and never from somatic cells within the gonad. The only stage in the development of meiotic cells (oocytes, spermatocytes) when labelling with tritiated thymidine can be carried out is the pre-leptotene stage. In the ovaries of most species this occurs only before birth, but in the vole, mink, ferret, rabbit and golden hamster the process occurs shortly after birth. Oocytes in the ovaries of mature mammals are only labelled if the isotope is injected at the pre-leptotene stage, thus confirming that the definitive germ cells in the female are the direct progeny of the stem cells and consequently of the primordial germ cells.

Similar studies have shown the continuity of male germ cells from embryo to adult.

In conclusion it should be pointed out that there is a small group of female mammals in which early germ cells persist in the mature ovary as oogonia and oocytes at early stages of meiotic prophase. These include various primitive primates, such as the bush-baby, *Galago*; the lorises, *Loris* and *Nycticebus*; and the aye-aye, *Daubentonia*. In these species the early sex-cell stages occur in groups, usually within epithelial cords in the cortex, while the definitive germ cells resemble those in other mammals (oocytes at the diplotene stage of meiotic prophase in primordial follicles; see Chapter 2). Injection of tritiated thymidine is followed by labelling of the germ cells within the 'nests' but never of the definitive oocytes. The epithelial cords and their germ cells are thus seen merely to be an embryological curiosity which in other mammalian species would have been eliminated before, or shortly after, birth. Occasional cells resembling primordial germ cells can be found at the hilum of the ovary, especially in the cat and Striped skunk, but the development of these cells (like those in the 'nests') beyond a primitive stage would appear to be a rare event. There is no evidence that these germ cells are derived from any source other than the embryonic primordial germ cells and certainly not from the 'germinal' (coelomic) epithelium at any stage during the life of the individual.

Prodigality is the keynote in the early history of the germ cells. Even in the migratory phase of this population, losses are large, and later after the colonization of the gonads a more drastic elimination occurs. The explanation could invoke selection in some way, but at the present time we can only speculate, as we must also concerning the origins and lineage of these most important mammalian cells.

SUGGESTED FURTHER READING

Observations on the movements of the living primordial germ cells in the mouse. R. J. Blandau, B. J. White and R. E. Rumery. *Fertility and Sterility* **14**, 482 (1963).

The development of genital glands and the origin of germ cells in human embryogenesis. L. I. Falin. *Acta anatomica* **72**, 195 (1969).

The development of the gonads in man, with a consideration of the role of fetal endocrines and the histogenesis of ovarian tumours. J. Gillman. *Contributions to Embryology* **32**, 83 (1948).

Migration of the germ cells of human embryos from the yolk sac to the primitive gonadal folds. E. Witschi. *Contributions to Embryology* **32**, 69 (1948).

Marshall's Physiology of Reproduction. Ed. A. S. Parkes, (especially chapter 4, vol. 1, part 1), 3rd edition. London; Longmans (1956).

Reproductive Physiology of Vertebrates. A. van Tienhoven. Philadelphia; W. B. Saunders (1968).

Sex and Internal Secretions. Ed. W. C. Young. 3rd edition, 2 vols. London; Baillière, Tindall and Cox (1961).

The Ovary. Ed. S. Zuckerman, (especially chapter 1), 2 vols. London; Academic Press (1962).

2 Oogenesis and ovulation

T. G. Baker

Oogenesis is best defined as the formation, development and maturation of the female gamete. As we saw in Chapter 1, the processes involved begin in embryonic life and continue to the time of ovulation. By far the most important is the process of meiosis, which is in a sense the antithesis of fertilization, the former halving the number of chromosomes while the latter restores the diploid complement.

Ovulation is the process whereby the egg is released from the ovary at a time appropriate for its fertilization and further development. Ovulation involves rupture of the Graafian follicle which has been developed by follicular growth under the control of gonadotrophic hormones from the pituitary gland. The process of ovulation is dependent on the correct endocrine balance to permit full follicular growth to occur. As we shall see later, the final stages of the meiotic divisions are dependent on the completion of follicular growth and the release of pituitary hormones.

MEIOTIC CHANGES IN OOCYTES

Meiosis consists of two cell divisions, the first of which involves the halving of the chromosomal number (diploid to haploid) and permits the exchange of genetic information between pairs of chromosomes, one member of each pair having been derived from the mother and the other from the father. The second division resembles mitosis, although only the haploid chromosomes are involved. In the human female, all the cells of the body (including the oogonia) undergo mitotic divisions, the number of chromosomes remaining constant at 46. Primary oocytes are germ cells in which the first meiotic division occurs, resulting in secondary oocytes with only 23 chromosomes. Each chromosome is split longitudinally into a pair of chromatids which separate during the second meiotic division, so that the egg is left with single chromosomal threads (see Fig. 2-1). As we shall see later, both of these nuclear divisions are accompanied by unequal cytoplasmic divisions, so that two small polar bodies are formed.

Cell division – whether mitotic or meiotic – proceeds in turn through prophase, metaphase, anaphase and telophase (Fig. 2-1). The interval between successive cell cycles is termed interphase, and is important since it is then that DNA replication occurs. The terms used for subdividing cell division are purely arbitrary, however, since the process is dynamic and proceeds with few interruptions once started. The prophase of the first meiotic division is exceptional in that it is greatly prolonged in time, is highly specialized, and in the female is interrupted by two periods of arrested development (the dictyate stage and metaphase I or II, to which we will return later). This prophase can conveniently be divided into five stages identified by cytologists as leptotene, zygotene, pachytene, diplotene and diakinesis (the so-called 'germinal vesicle' stage).

Shortly after the last mitotic division by the oogonium (see Chapter 1), the cell enters an interphase in which DNA is replicated in preparation for meiosis. This is the only time in the first meiotic division that the germ cells incorporate such radioactive precursors of DNA as tritiated thymidine. During this stage (which is often called pre-leptotene) the nucleus acquires a speckled appearance and short segments of chromosomal threads may be seen. The cell enlarges and shows an increased affinity for nuclear dyes until eventually the diploid number of chromosomes can be identified in cytological preparations (leptotene stage; see Fig. 2-1). Let us consider a hypothetical example in which the germ cells contain four chromosomes, as shown in the illustration. There are two long chromosomes and two short chromosomes; one of each type is derived from the mother, the other from the father. These are

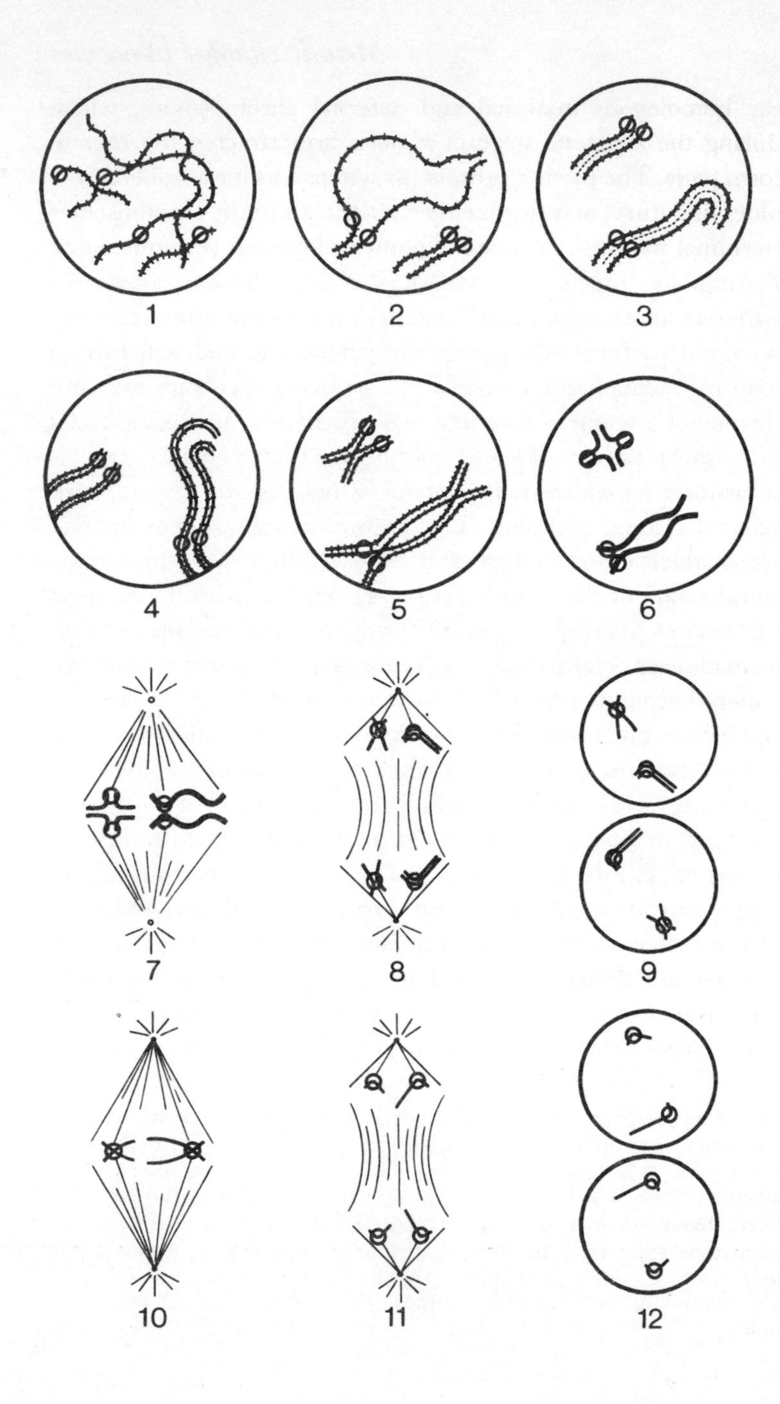
1
2
3
4
5
6
7
8
9
10
11
12

the homologous maternal and paternal chromosomes, which during the zygotene stage of meiosis are attracted together to form pairs. The pairing process (or synapsis as it was called in the older literature) may begin either at the ends of the chromosomes (terminal pairing), or at some point in between (intermediate). Pairing is so precise that homologous genes become associated with one another; if a small region of one chromosome (say the maternal partner) is missing, the paternal thread will form a loop to maintain the correct gene sequence (see Fig. 2-2) but this is not a feature of normal meiosis. Pairing is accomplished in a short period of time and hence the zygotene stage is transitory. In contrast, pachytene, which rapidly ensues, is of relatively long duration. The chromosomes are now paired along their entire length and they shorten and thicken by spiralization of the threads (Fig. 2-1). Studies with the electron microscope have shown that the 'synaptonemal complexes' thus formed each consist of two outer threads of chromosomal material separated by a fine association line. As the cell grows and gains a great complexity of cytoplasmic organelles (seemingly the 'late' pachytene stage), the lateral arms of the synaptonemal complex split longitudinally to form four threads arranged in pairs (the complexes now being termed tetrads or bivalents), while the intermediate line remains unchanged. These threads break and rejoin during spiralization, and parts of the maternal and paternal homologues are exchanged. The changes are difficult to detect in cytological preparations and are inferred in retrospect from studies of oocytes at the subsequent stage, diplotene. The chromosomes now show mutual

Fig. 2-1. Diagram showing the behaviour of chromosomes during meiosis. For simplicity only two pairs of chromosomes are depicted, a short pair with terminal pairing, and a long set with intermediate synapsis. Only the nuclei are shown; the cytoplasm of the cells has been disregarded. 1, leptotene; 2, zygotene; 3, pachytene; 4, late pachytene showing tetrad (two pairs of bivalents); 5, diplotene; 6, diakinesis; 7, metaphase I; 8, late anaphase I; 9, telophase I (idealized); 10, metaphase II; 11, anaphase II; 12, telophase II (ovum and polar body).

repulsion from each other so that the bivalents separate except at certain points (chiasmata; Figs. 2-1 and 2-3) where 'crossing-over' has occurred during the phase of chromosomal breakage and reunion. The exchange of genetic material between maternal and paternal threads results in a reassortment of genes, which ensures that the chromosomes of the oocyte are different from those of either parent. The homologous chromatids constituting each bivalent still remain attracted to each other, and hence the pattern characteristic of the diplotene stage is maintained (see Fig. 2-3).

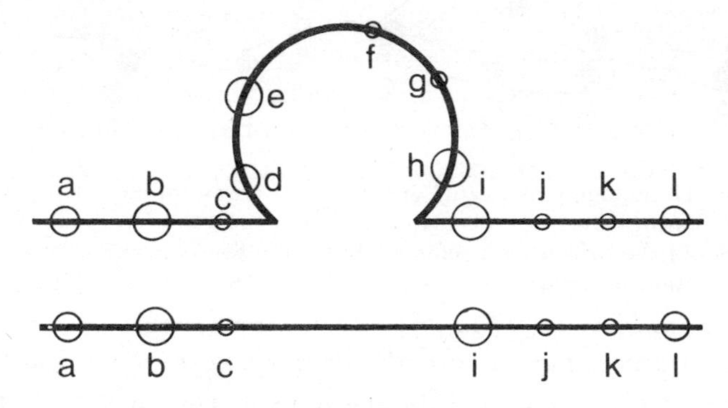

Fig. 2-2. Diagram showing precision of pairing at zygotene. Individual genes on maternal and paternal chromosomes are paired, apart from '*d–h*' which are missing from one homologue.

The description of meiotic prophase so far applies equally well to oogenesis and spermatogenesis, but now the similarities end. Meiotic prophase in the female is completed to the diplotene stage shortly after birth in most species, and the cell enlarges greatly. Oocytes then enter a prolonged 'resting phase' which is terminated shortly before ovulation with pre-ovulatory meiotic changes in the Graafian follicle. In species such as man, the earliest viable oocyte to resume meiosis does so at the time of puberty, and the last egg to pass through preovulatory maturation may be found in women aged 45–50 years. That this long period of arrested development may be of great importance is suggested by the observation that the incidence of embryos with meiotic defects (e.g. the extra chromosome number 21 in mongolism) increases with maternal age, but there are other possible explanations for these defects (see Book 2, Chapter 5).

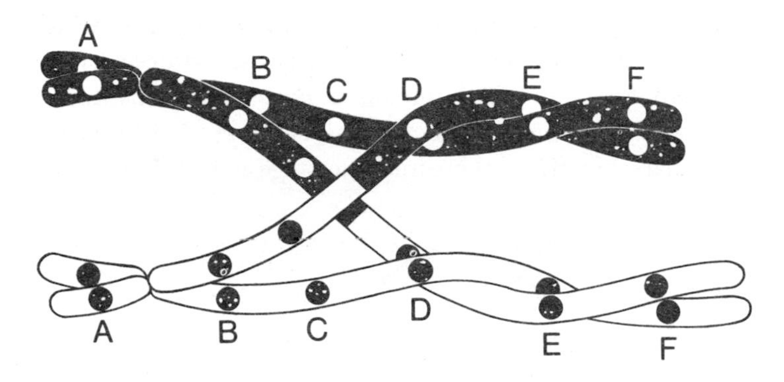

Fig. 2-3. Homologous maternal and paternal chromosomes at diplotene. 'Crossing-over' has occurred between one chromatid of each pair; these have broken and rejoined, and have thus exchanged parts. (*A–F*, positions of genes.)

By contrast, meiosis in the male does not begin until around the time of puberty, but then continues without interruption throughout adult life. There is no period of arrested development in the spermatocyte, diakinesis following immediately after the diplotene stage (see Chapter 3).

These factors have a profound effect on the number of germ cells available for reproduction. In the male, the number of spermatozoa that can be produced is vast since the stem cells, or spermatogonia, are being continually replaced throughout life by mitosis. As we saw in Chapter 1, the stem cells of the ovary have a limited life and probably a finite number of mitotic divisions. Consequently, the population of germ cells in the female has a fixed upper limit which, with the disappearance of oogonia, is rapidly depleted with increasing age by the process of atresia. By way of example we may refer again to the situation in the human female as described in Chapter 1: the number of germ cells increases from about 1700 during migration, to some

600 000 during the 2nd month of gestation, and subsequently to almost seven million by mid-term (see Fig. 2-4). The population of germ cells then declines rapidly to about two million (of which half are already degenerate) at the time of birth. This decline, as we shall see later in this chapter (p. 41), is due mainly to the elimination of large numbers of cells by the

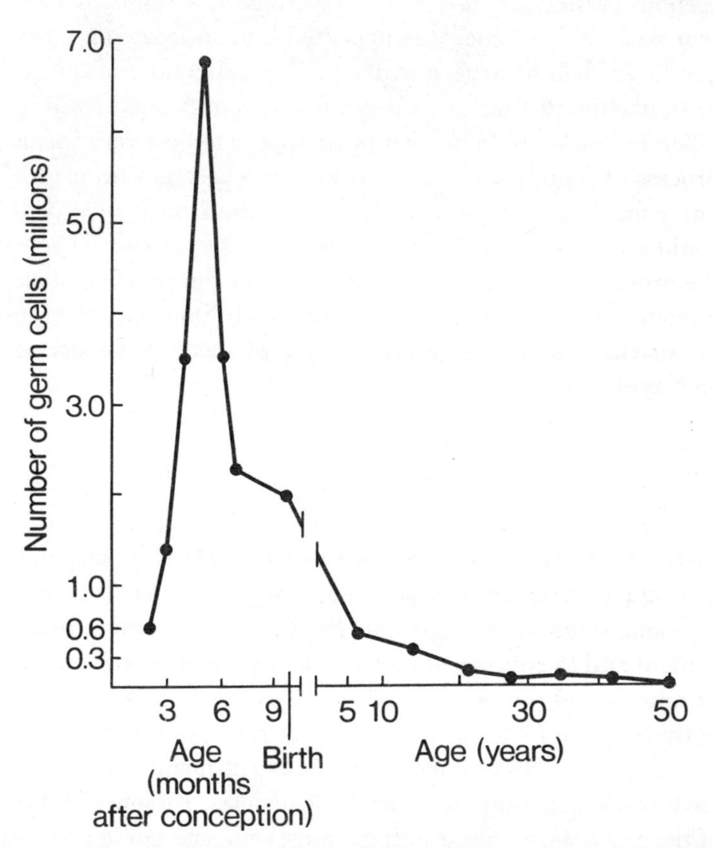

Fig. 2-4. Changes in the total population of germ cells in the human ovary with increasing age. (From T. G. Baker, Radiosensitivity of mammalian oocytes with particular reference to the human female, *Amer. J. Obstet. Gynec.* **110,** 746–61, Fig. 1 (1971); data for prenatal period derived from T. G. Baker, *Proc. Roy. Soc., Biol.* **158,** 417 (1963); data for children and adults from E. Block, *Acta Anat.* **14,** 108 (1952).)

process of atresia. But a contributing factor is the eventual cessation of mitosis in oogonia and their transformation into oocytes at the leptotene stage of meiosis. The number of oocytes continues to decline with increasing age until the time of the menopause in women (or the cessation of reproductive function in animals), when few oocytes can be detected in histological sections of the ovary (see Fig. 2-4). Of the seven million oocytes that were at one time present in the human ovary only about 400 to 500 will have been ovulated. The remainder of the germ cells, like the vast majority of spermatozoa in the male, will have fallen by the wayside in what must appear to be a very wasteful process of reproduction. This conclusion becomes firmer when one considers that the number of children born to a couple could hardly exceed twenty. As we shall see in later chapters, the process is not quite so wasteful as these figures suggest since numerous other factors are involved which limit both the number of viable offspring and the timing of successful conception and implantation.

The dictyate, diffuse diplotene, or dictyotene stage of meiotic prophase

In the rat the process of meiosis in the ovary is synchronized, and 90 per cent of the germ cells at any particular time are at the same stage of development. In other species, asynchrony is evident and the ovaries of fetuses contain germ cells at all stages of mitosis and meiosis.

By the time of birth, the ovaries of rats, guinea pigs, sheep, cows, monkeys and human beings contain mainly oocytes that have reached the diplotene stage of meiosis. In contrast, those of the newly born rabbit, ferret, mink, vole and golden hamster contain only oogonia, the prophase of meiosis being completed within the first few weeks after birth. The process of oogenesis in the pig occurs mainly in fetal life but extends into the post-natal period, while in the cat oocytes at early stages of meiotic prophase may be found up to the time of puberty. Inbreeding can affect the temporal relations of oocytes: CBA and A strains of laboratory mice contain oocytes that have reached the diplotene stage at birth, while those of the Street and Bagg strains are mainly at pachytene.

In all these species, irrespective of the timing of oogenesis, oocytes embark on a prolonged 'resting' phase shortly after the onset of diplotene. This so-called dictyate or dictyotene stage is characterized by highly diffuse chromosomes, the DNA of which has little affinity for such nuclear stains as Feulgen's reagent. In the extreme condition, found in oocytes of rats and mice, the nucleus contains what seems superficially to be a reticulum of fine threads. Some authors believed that these threads 'disappeared', to be reconstituted at a later date from within the nuclear sap, and also (largely on the basis of inadequate or imprecise histochemical techniques) that the oocyte contained virtually no RNA and thus lacked the ability to manufacture the protein and other materials that can loosely be grouped as 'yolky substances'. For these reasons the period of arrested development was considered to be truly a resting phase in which the oocyte was merely 'nursed' by its investing follicle cells, its growth being only passive.

There can be no doubt that oocytes at the dictyate stage grow considerably, before as well as after the onset of follicular growth (Fig. 2-5). But follicle cells are not necessary for the growth of the oocyte since their absence is occasionally observed following irradiation or hypophysectomy. Moreover 'giant oocytes' whose granulosa envelope remains undeveloped have been reported for numerous species including mouse, human and Rhesus monkey.

Studies with the electron microscope have shown that both the oocyte *and* its granulosa cells contain the cytoplasmic organelles that are usually associated with cells secreting proteins and mucopolysaccharides. Thus oocytes contain Golgi vesicles, endoplasmic reticulum (mainly of the smooth type), and abundant rosettes of ribosomes as well as occasional vesicles of what could conceivably be the products or precursors of secretion. Unfortunately, the use of radioactive tracer substances has so far proved of little value in assessing the roles of the oocyte and granulosa cells in the growth phase, since the precursors involved are incorporated by almost all cells within the ovary, including oocytes.

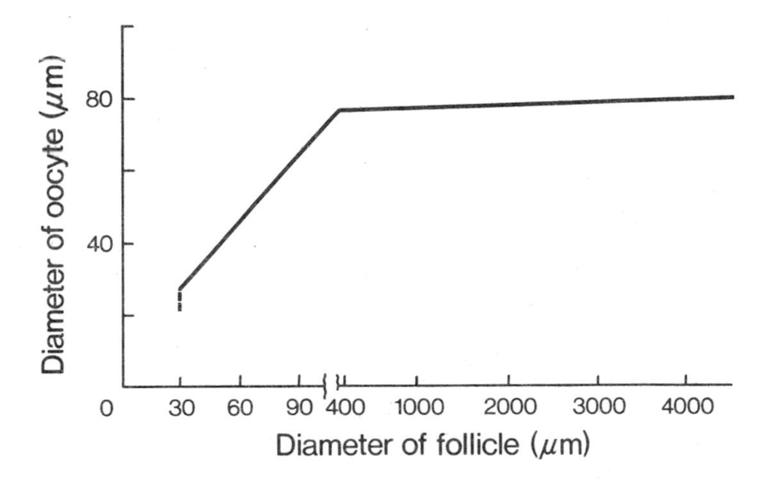

Fig. 2.5. Growth of the oocyte and follicle in the human ovary. First the oocyte alone enlarges, then both grow at a corresponding pace; finally growth is restricted almost entirely to the follicle. (Modified from S. H. Green and S. Zuckerman, *J. Anat., Lond.* **85**, 373, Fig. 1 (1951).)

The results of recent studies on oocytes at the dictyate stage have shown that the chromosomes bear lateral projections in the form of branches and loops (Fig. 2-6) which actively replicate ribonucleic acid (RNA). They closely resemble the lampbrush chromosomes which are found almost universally in eggs of lower vertebrates and some invertebrates. The RNA may act as the messenger directing protein synthesis within the oocyte itself. Circumstantial evidence from studies of oocytes in frogs and toads suggests that some of the RNA may act as the early 'organizer' of mammalian development. Whatever the outcome of the additional studies required to elucidate these complex problems, the dictyate stage should clearly not be referred to as a resting phase, since the oocytes show a high degree of metabolic and synthetic activity at a time when the follicular envelope consists of only a few flattened epithelial

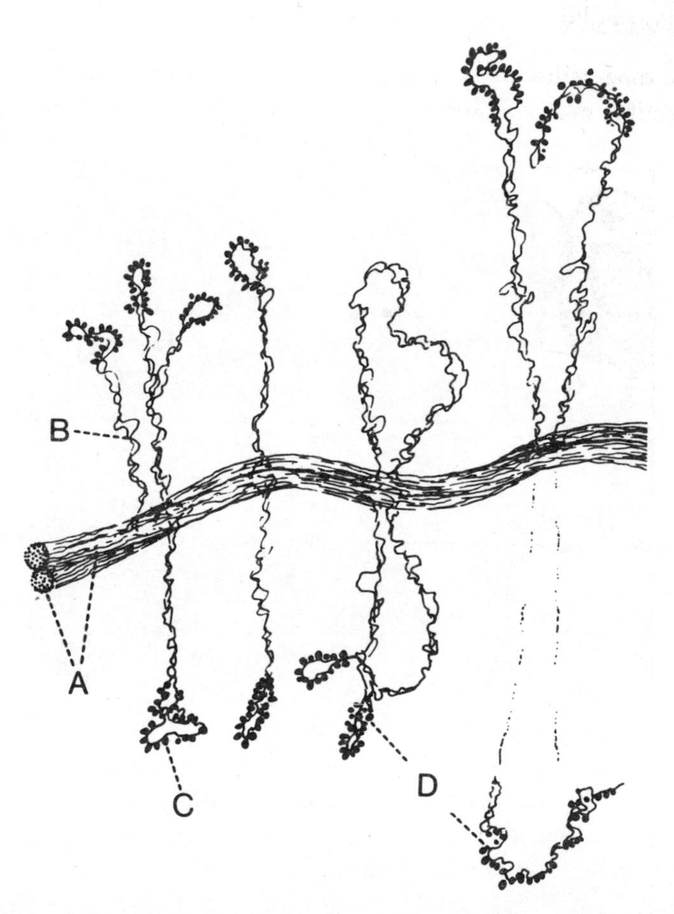

Fig. 2-6. Diagrammatic representation of a chromosome at the diplotene/dictyate stage of meiotic prophase. Fibrils (B) constituting loops and lateral projections arise from the bivalent (A). Granular material containing ribonucleoprotein is associated with the ends of the projections (C) and loops (D). (T. G. Baker and L. L. Franchi, The structure of the chromosomes in human primordial oocytes, *Chromosoma*, **22,** 358–77, Fig. 14. Berlin – Heidelberg – New York; Springer (1967).)

cells. Only during the second half of the oocyte's growth phase do the granulosa cells clearly contribute maternal protein and other materials to the ooplasm (see later).

FORMATION AND FUNCTION OF THE ZONA PELLUCIDA

The zona pellucida consists of mucopolysaccharide and trypsin-digestible material, and is indistinct in sectioned material unless

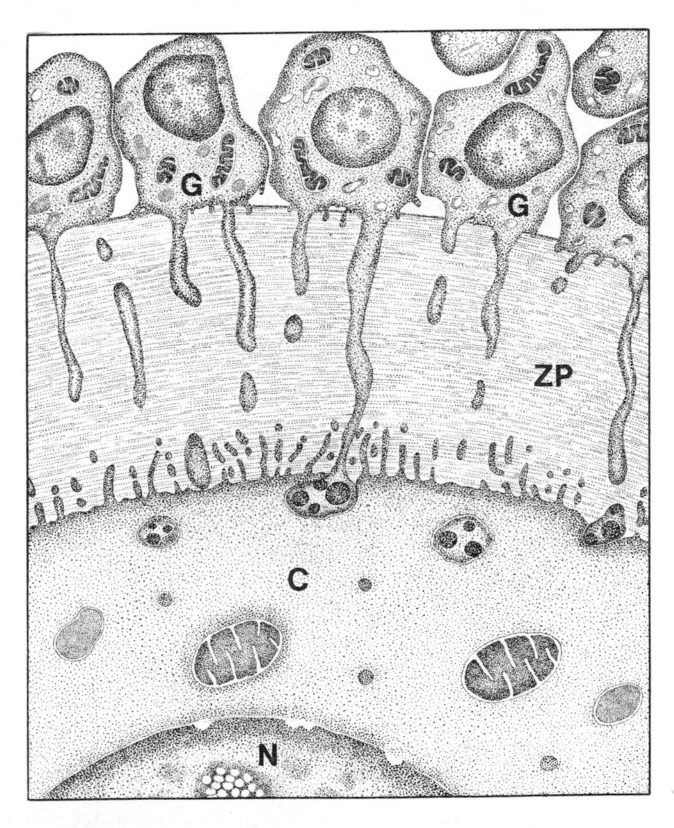

Fig. 2-7. Structure of fully formed zona pellucida (ZP) around an oocyte in a Graafian follicle. Microvilli arising from the oocyte interdigitate with processes from the granulosa cells (G). These processes penetrate into the cytoplasm of the oocyte (C) and may provide nutrients and maternal protein. (N, oocyte nucleus.)

stained by specific histochemical stains (as in the periodic acid–Schiff, or PAS reaction). When viewed with the electron microscope it appears somewhat 'fluffy' due to precipitation of its protein component with the fixatives employed (see Fig. 2-7).

The zona pellucida forms around oocytes that are surrounded by a complete layer of cuboidal granulosa cells. Islands of fibrillar material are deposited in spaces between adjacent granulosa cells and the oocyte surface (Fig. 2-8). The source of the fibrils is obscure although islands of material within the endoplasmic reticulum of both the follicle cells and the oocyte may represent precursor substances. The cellular transformation of proteinaceous substances into mucopolysaccharides is

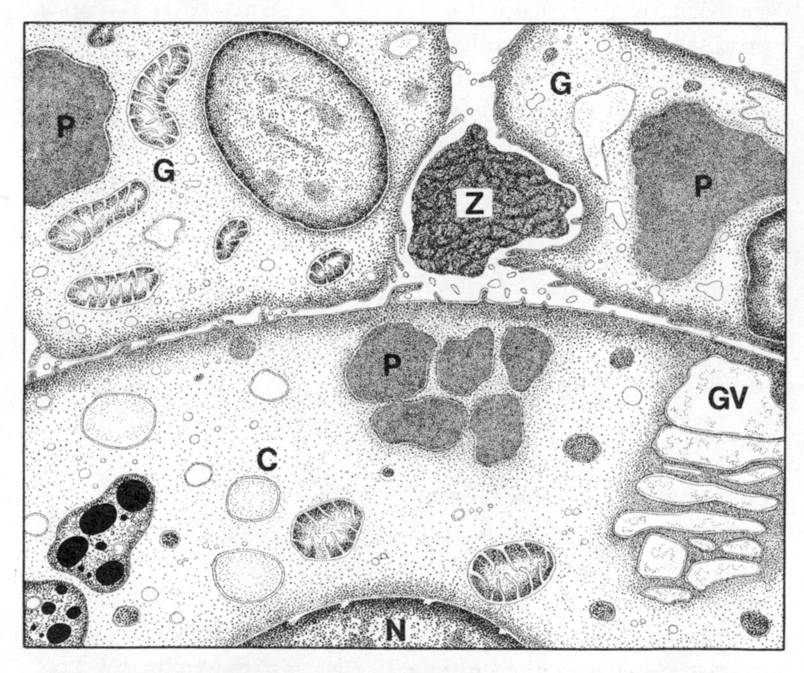

Fig. 2-8. Idealized electron micrograph depicting initial formation of the zona pellucida (Z) between granulosa cells (G). Precursor-like material (P) is found in the granulosa cells and the oocyte; the Golgi vesicles (GV) in the oocyte may have been responsible for the formation of the oocyte's share of this material. (N and C are the nucleus and cytoplasm of the oocyte.)

known to occur in Golgi vesicles, which are usually found close to the oocyte surface in regions where the zona is being deposited. Furthermore, the material of the zona pellucida is not homogeneous but consists of two layers, the outer staining more intensely than the inner, and so the zona pellucida may well be the product of both oocyte and its follicular envelope. There can be little doubt as to the role of the granulosa cells in the processes involved but further studies are required to determine the part played by the oocyte.

The islands of zona material (Fig. 2-7) fuse together and eventually form an impervious jelly-like coat which completely surrounds the egg. The structure in some ways resembles a sieve, in that microvilli from the oocyte surface and coarser processes from the granulosa cells occupy canals within the zona (Fig. 2-8). The granulosa-cell processes often traverse some distance into the oocyte and may transfer maternal protein into vesicles that are budded off into the cytoplasm. The zona is impervious to solutes of high molecular weight such as polysaccharides and proteins. The granulosa cells, which later become transformed into the cumulus oophorus, are clearly essential for the nutrition of the egg once the zona pellucida is established; they produce essential substances such as lactic and pyruvic acids.

FOLLICULAR GROWTH

Soon after oocytes are formed, they become surrounded by a single layer of flattened epithelial cells, and the primordial follicle is established. Follicular growth involves a change in shape of the cells, which become cuboidal in form, and an increase in number by mitosis. Fluid subsequently accumulates in spaces between the epithelial cells, and the follicle is now described as being vesicular.

The ovary is a dynamic structure in which vesicular follicles are constantly developing from those of the primordial type. The first 'growing' follicles appear in the ovaries within a few days of birth in most species, although in primates they may occur before birth. But only with the establishment of the correct hormonal balance and reproductive cycles at the time of puberty is the process of follicular growth permitted to culminate in ovulation. (The hormonal control of follicular growth and ovulation is dealt with more fully in Book 3.) The great majority of vesicular follicles that are produced (and all of those before puberty) undergo degeneration at varying stages in their formation (see p. 43). The number of follicles that attain ovulation is more or less fixed for the species by the levels of circulating gonadotrophic hormones, a point that we shall return to later. Thus in the human female only one follicle usually undergoes ovulation each month, the remaining twenty or so that had reached the same stage of growth degenerate. The injection of additional hormones causes more follicles to ovulate, a technique that has applications in agriculture and has inadvertently resulted in multiple births in women treated with gonadotrophins.

About 90 per cent of all oocytes in the ovaries of sexually mature mice, rats and monkeys are enclosed within primordial follicles consisting of one layer of flattened granulosa (epithelial) cells. This follicular envelope is initially incomplete, the cells being scattered about the periphery of the oocyte. The remaining 10 per cent are the 'growing' and Graafian follicles which are often classified for convenience by histological criteria, such as: (i) size; (ii) numbers of granulosa cells (or layers of such cells) within the membrana granulosa; (iii) development of the theca, and (iv) position of the oocyte within its surrounding cumulus oophorus. The schemes devised for identifying follicular stage vary in their complexity, that shown in Fig. 2-9 being a simplified type that can be used for a wide range of species.

The earliest signs of follicular growth in primordial follicles are, (i) an increase in the size of the oocyte, (ii) a change in shape of the granulosa cells from flat to cuboidal, after which they multiply by mitosis, and (iii) the formation of the zona pellucida.

Subsequently, by further division of granulosa cells, the membrana granulosa becomes two, three and then four layered. At this time blood capillaries invade the fibrous layer of cells surrounding the follicle and form a vascular layer, the theca interna. This is surrounded by fibroblasts of the theca externa and is the only source of nutrients both to the membrana granulosa and the oocyte.

During each reproductive cycle, as a direct consequence of the release of sufficient FSH by the pituitary gland, a crop of 'growing' follicles (stage 5 in Fig. 2-9) are stimulated to undergo further growth and maturation. The number of follicles that are

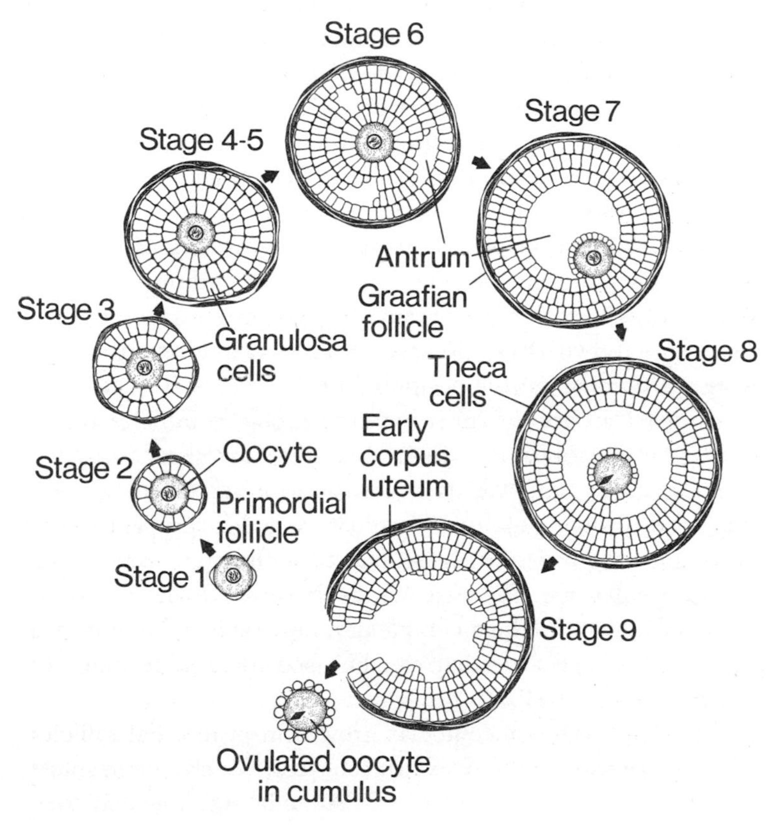

Fig. 2-9. Diagrammatic representation of follicular growth.

thus 'selected' is determined by the available quantity of gonadotrophic hormone. The pituitary-dependent phase of follicular growth involves further multiplication in the number of granulosa cells, and also the passage of fluid ('liquor folliculi') into spaces between them (see stage 6 in Fig. 2-9). This fluid resembles blood serum and is probably derived directly from the capillaries with only slight modification by the follicle cells. As the quantity of fluid increases, the cavities that it occupies increase in size and become confluent to form the antrum. The follicle is now said to be of the Graafian type, after Regnier de Graaf (1641–73) who first described them adequately. With the further expansion of the antrum, the oocyte occupies a position at one side of the follicle and is surrounded by two or more layers of granulosa cells. The innermost layer of cells becomes columnar in shape and constitutes the corona radiata or cumulus oophorus which persists around the egg for a period after ovulation. The dissolution of these cells whilst the egg is still in the follicle or shortly after its departure, is a sure sign that degenerative changes are occurring which will result in the death of the oocyte.

Our knowledge of the timing of stages in follicular growth has recently been advanced from studies of the incorporation of tritiated thymidine into granulosa cells. With each cell division, the number of silver grains over the cells in autoradiographic preparations is reduced by a half. Hence the transit times for each follicular stage can be determined from grain counts and used to estimate fluctuations in the population at each stage of growth. Such studies have shown that follicular growth is a continuous process and that the number of follicles at each of the stages from 1 to 5 fluctuates only slightly with the phases of the reproductive cycle. The populations are thus in a 'steady state', and follicles progress to the subsequent stage only to maintain the number at that stage. But with the onset of stage 5, the numbers of follicles fluctuate widely, being most common during the early part of the cycle and least common after ovulation.

The factors initiating follicular growth in primordial follicles remain completely unknown at present. It is difficult to explain how a few such follicles are 'selected' to embark on the growth phase, while seemingly identical neighbouring follicles remain unaffected. The earliest signs of growth almost certainly occur in the oocyte itself which may trigger the changes in granulosa cells. The oocyte may produce a 'messenger' (possibly RNA from lampbrush chromosomes, as described earlier) at a specific time in its growth phase. Alternatively, the oocytes may be 'programmed' in some way. Perhaps the time of onset of meiotic prophase in germ cells during embryonic or fetal life affects the timing of their subsequent growth phase. The first cells to enter meiosis would also be the earliest to embark on follicular growth. This 'production-line' hypothesis may in part account for germinal selection in the ovaries of cows, monkeys and women in which meiosis is not synchronized and occupies a period of months in the life of the fetus; but it can be of little value in other species where meiotic prophase is completed in synchrony within at most a few days (e.g. mouse and rat). Pituitary hormones are unlikely to initiate follicular growth since the initial phases of the growth process continue after removal of the pituitary gland. Growth up to four layers of granulosa cells is generally held to be independent of hormonal control, while the process beyond this stage is hormone dependent (see below). Paradoxically, the injection of antibodies to gonadotrophic hormones into mice during the first 2 weeks of life is said to disturb the normal pattern of development in granulosa and thecal cells, but does not affect growth by the oocyte. Clearly the regulation of the process of follicular growth is highly complex and requires the elaboration of new techniques if the mechanism is to be further resolved.

In contrast, our knowledge of the development of the Graafian follicles from the 'precursor pool' (stages 6 to 8 in Fig. 2-9) has been considerably enhanced from studies of ovaries in organ culture, and from those involving replacement therapy in animals whose pituitary glands have been surgically removed.

In both situations, the injection of follicle-stimulating hormone preparations from pregnant mares' serum and from pituitary glands, and of natural and synthetic oestrogens, can promote further growth of the follicle and antrum formation. Relatively pure preparations of pituitary follicle-stimulating hormone (FSH) are said to be less effective than those containing luteinizing hormone (LH); the optimum concentration may be about 1:3 respectively. It is also known that the action of mares' serum gonadotrophin (PMSG) on follicular growth in hypophysectomized rats is potentiated by pretreatment with oestrogens. It would seem that the principal controlling mechanism in the final growth phase of the follicle and its antrum involves a peak of FSH in the presence of some LH, the function of which may be to ensure adequate steroid synthesis (especially of oestrogens). The combined effects of these hormones determines not only the *number* of follicles that develop to the mature Graafian type, but also the *proportion* that undergo ovulation compared with those that degenerate.

PRE-OVULATORY MATURATION

If ovulation is to occur, the Graafian follicle (and the egg within it) must undergo further changes that can only take place under precisely controlled hormonal conditions. The level of circulating FSH remains elevated for only a short time at the beginning of the growth phase, after which the quantity of circulating gonadotrophin (FSH and LH) remains fairly constant until a short time before the impending ovulation. The interval between the LH peak and ovulation varies from about 12–15 hours in the mouse, rat and rabbit, through 24 hours in the pig, to some 36 hours in the human female. The onset of pre-ovulatory maturation is marked by a sudden and dramatic rise in the release of gonadotrophins from the pituitary, especially of LH (the so-called 'LH surge'). In reality, both the FSH and LH levels rise and consequently the ratio of these hormones one to the other is probably just as important in controlling pre-ovulatory maturation as that of LH alone. In any event, these hormones reach an optimum value for the species which affects the final maturation of the oocyte and its surrounding follicle; they also regulate to some extent the number of eggs that ovulate, which is more or less constant for each species.

The LH surge (or the increase in oestrogen which causes it) is said to induce a final 'wave' of mitosis in granulosa cells so that their number reaches an optimum size for the ovulatory Graafian follicle. The quantity of follicular fluid in the antrum also increases dramatically; due to increased permeability of the blood–follicle barrier, the intra-follicular pressure initially rises in response to LH and then fluctuates before ovulation. The follicle thus enlarges greatly, its final size varying between species (e.g. bat 0.3 mm; rat 1.5 mm; pig 8 mm; man 10–15 mm; horse 30–50 mm).

During this final maturation of the follicle the cells adjacent to the egg acquire their characteristic columnar shape. Initially the cumulus oophorus is attached to the membrana granulosa over a wide area (see Fig. 2-9). This zone gradually becomes reduced in extent, partly by movements and loosening of the granulosa cells but also by fluid accumulation. The area of attachment is eventually reduced to a small stalk, by which time the innermost layer of cumulus cells (now called the corona radiata) consists of long thin cells with nuclei distal to the egg. The oocyte with its investments of coronal cells may become free-floating in the follicular fluid shortly before ovulation.

These changes in the structure of the Graafian follicle are accompanied by a resumption of meiosis within the oocyte. At the onset of pre-ovulatory maturation the egg is still a primary oocyte whose progression through meiosis was interrupted by a prolonged period of arrested development immediately following diplotene. At the time of the LH surge the oocyte contains lampbrush chromosomes, but the synthesis of RNA that is characteristic of early stages of growth has now terminated. The chromosomes subsequently shorten and thicken, and their lampbrush loops (Fig. 2-6) are withdrawn. Chiasmata move along the chromosomes to become terminal and the threads resemble 'crosses' and 'chains' within the nucleus (see Fig. 2-1). These events occur at what is often called the 'germinal vesicle' stage of meiosis (more correctly referred to by cytogeneticists as diakinesis), and complete the prophase of the first meiotic division. Metaphase I rapidly ensues in which the bivalents arrange themselves on (and become attached to) microtubules at the equator of the meiotic spindle. During anaphase the bivalents move to opposite ends of the spindle, which rotates gradually through 90°, so that its axis becomes radially oriented (see Figs. 2-1 and 2-10). The rotation is completed by early telophase when the surface of the egg near to the chromosomes may form a slight elevation. The repulsion of the two sets of chromosomes is now complete and the area of the spindle between them is elongated. Division of the cytoplasm of the oocyte rapidly occurs as a furrow around the outermost set of chromosomes. This division of the egg is unequal (non-equational) in that one daughter cell – the secondary oocyte – receives most of the cytoplasm while the other contains the minimum of ooplasm and is the abortive polar body (see Figs. 2–10 and 2–11). In some mammals the polar body soon degenerates, but in others it persists for prolonged periods, even through the first few cleavage stages. The first polar body sometimes divides into two slightly smaller daughter cells at a time when the second meiotic division occurs in the egg.

Shortly after the extrusion of the first polar body the secondary oocyte embarks on the second meiotic division. Prophase is very short or seemingly non-existent, the chromosomes condense to form a half-moon-shaped mass at the periphery of the oocyte (chromatin mass stage). Spindle fibres subsequently appear adjacent to the chromatin and the chromosomes arrange themselves on the metaphase plate. Metaphase II in the majority of species is a stage of arrested development like the dictyate stage, and it is now that ovulation generally occurs. The subsequent meiotic maturation of the oocyte is dependent on penetration by a spermatozoon at fertilization, which is the subject of Chapter 5.

There can be no doubt that the resumption of meiotic

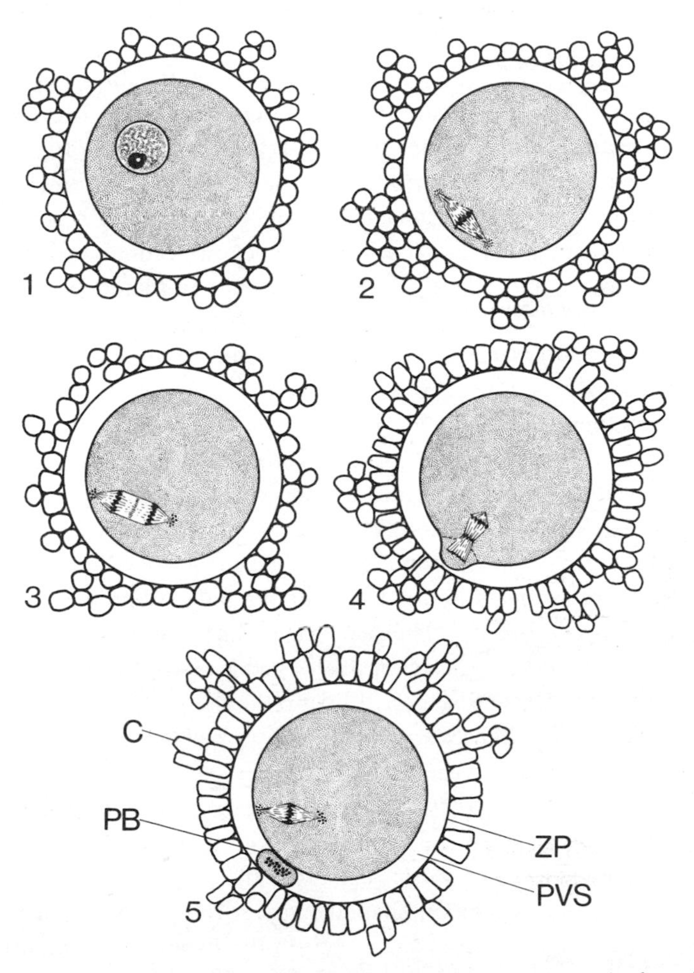

Fig. 2-10. Pre-ovulatory maturation of the egg. 1, germinal vesicle stage; 2, metaphase I; 3, anaphase I and rotation of the spindle; 4, polar body extrusion; 5, secondary oocyte at metaphase II, surrounded by a perivitelline space (PVS) between egg and zona pellucida (ZP). PB is the first polar body. C is one of the cells of the corona radiata.

changes within oocytes in Graafian follicles is induced by the LH surge, but the precise mechanism is open to dispute. If oocytes are removed from their follicles prior to the release of LH *in vivo* they resume meiosis spontaneously in a chemically

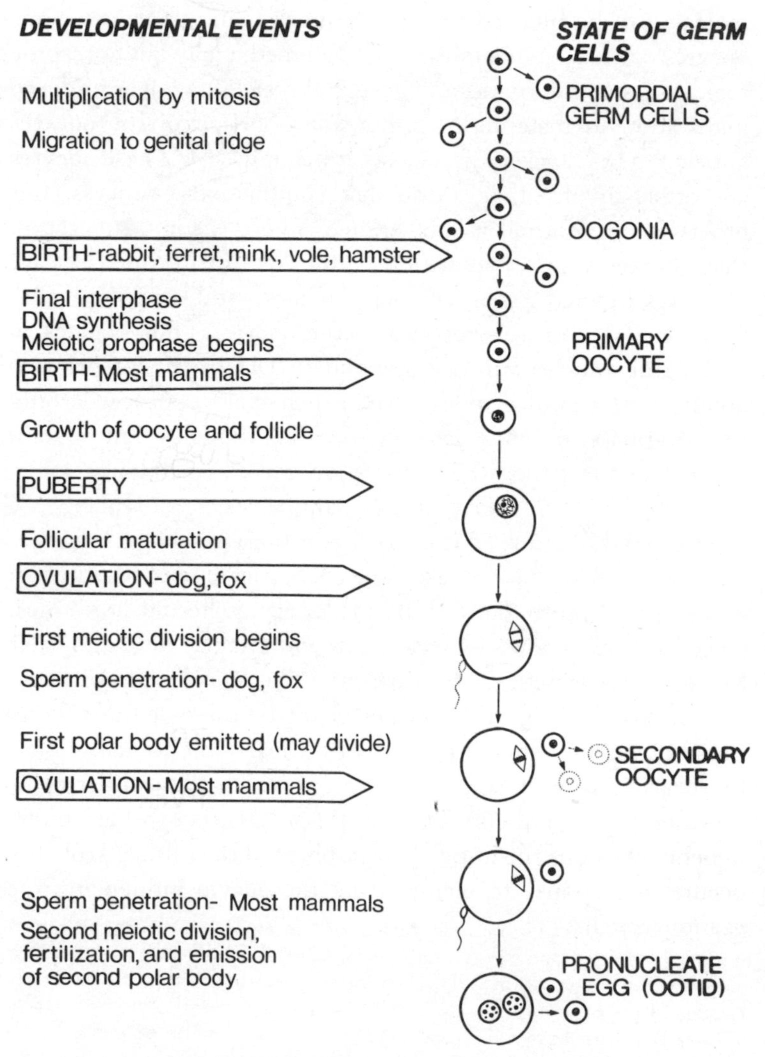

Fig. 2-11. Life-cycle of the female germ cell.

defined medium devoid of all hormones (both steroid and protein). It has therefore been suggested that the Graafian follicle in some way inhibits the maturation of the oocyte until it is 'triggered' by LH. The difficulty with this hypothesis is that at least two inhibitor substances would be required since meiosis can be blocked in mice at more than one stage. Thus oocytes cultured in chemically defined media may fail to resume meiosis altogether (persistent germinal vesicle stage), or undergo maturation to metaphase II. But a proportion of the oocytes appears to be 'blocked' in meiosis at metaphase I. These oocytes are often smaller than those that complete the process, but otherwise are normal in appearance. There is some suggestion that oocytes whose maturation is 'blocked' at metaphase I are usually obtained from smaller follicles, and therefore that follicular size controls meiosis in the oocyte. Thus oocytes in pre-antral follicles will not respond to LH treatment prior to about stage 5–6 of growth, after which meiosis proceeds only to metaphase I. Only eggs in Graafian follicles can sustain meiosis to the point where ovulation occurs (metaphase II).

The blockage of meiosis at the germinal vesicle and metaphase I stages could be due to immaturity of these cells or to changes in their metabolism. However, the change leading to resumption of meiosis is more likely to be mediated by steroid hormones, since these are known to accumulate in follicular fluid and their levels increase with LH treatment. The steroids presumably act on the cumulus cells rather than directly on the oocyte, possibly via metabolic intermediates. Clearly the follicle should be regarded as a dynamic structure in which the cellular components (granulosa, cumulus and theca cells; oocyte) are interdependent. Thus the initial growth phase of the follicle probably occurs in response to factors from the oocyte influencing the granulosa cells.

OVULATION

The oocyte is shed from the Graafian follicle by the process of ovulation at a precise time after the onset of the LH surge (see p. 32). The timing of ovulation in relation to the reproductive cycle varies considerably between species and may be dependent on diurnal rhythms, or induced by coitus (see Chapter 4).

During the final phase of pre-ovulatory maturation, a small area of the wall of the Graafian follicle (and the overlying ovarian cortex) becomes thin and translucent, forming the 'stigma'. Studies of sectioned follicles have shown that thinning follows pyknosis of an area of the membrana granulosa, the cells being phagocytosed or merely shed into the follicular cavity. Translucence of the stigma may be partly due to its becoming devoid of blood capillaries. Towards the end of the period in which these changes occur, the stigma becomes raised and resembles a blister on the surface of the ovary. With the onset of ovulation the blister tears open and the follicular fluid gently oozes out. The process is rarely explosive, the trickle of somewhat viscid follicular fluid continuing for a short time. Thus the egg, which by now is free-floating in the fluid, passes out passively with follicular debris (sloughed granulosa cells) and is directed towards the Fallopian tube by ciliary currents.

So much for the observed process of ovulation; but what of the mechanism responsible for these changes? There can be little doubt that injections of either LH or HCG (human chorionic gonadotrophin which has a luteinizing action) are effective in inducing ovulation, but the natural process *in vivo* probably involves a precise ratio of FSH:LH rather than LH alone. Several mechanisms have been proposed to account for the observed changes in the follicle wall, but most are easily discounted on physiological grounds. A combination of events may well occur, and considerable variation between species is to be expected.

Ovulation has often been attributed to increase in pressure of follicular fluid, owing either to continuous secretion or to contraction of the follicle by smooth muscle fibres and/or to pressure exerted by the oviduct. But ovulation is not an explosive phenomenon and manometric studies have shown that intra-follicular pressure merely oscillates about a value similar to that within capillaries. Thus pressure changes alone cannot account for the release of oocytes in most species, and injection of excess fluid into follicles of the pig does not induce ovulation.

The most widely accepted theory is based on the idea that the blister-like stigma results from ischaemia of capillaries. The blood flow into the ovary increases following treatment with LH and also with histamine, but is unaffected by FSH, prolactin and adrenalin. Obliteration of thecal capillaries in the region of the stigma is believed to be induced by the combined effects of high blood pressure and forces applied between the follicle and ovarian cortex. Necrotic changes in the membrana granulosa resulting in the translucence of the stigma could be due to the action of enzymes. Injections of pronase, collagenase or nagrase into rabbit follicles have been shown to induce ovulation in 62–100 per cent of Graafian follicles. Trypsin is less effective and chymotrypsin, hyaluronidase, lysozyme and saline solution are ineffective. Apparently the proteolytic enzymes are produced by follicular cells in which the proteins can be detected histochemically. Furthermore, injections of actinomycin D or puromycin (substances that block protein synthesis) into the ovaries of rabbits inhibit ovulation. The enzymes presumably have only a local effect since few granulosa cells become pyknotic. Intra-follicular pressure initially rises with enzyme treatment but then subsides to normal levels before ovulation occurs.

The egg at ovulation

Ovulation occurs in most mammals when the oocyte has reached the metaphase of the second meiotic division. There are, however, exceptions to this general rule. Thus in the fox and dog ovulation occurs around metaphase I and the first polar body is extruded shortly thereafter. By contrast the second meiotic division is said to be completed prior to ovulation in certain primitive insectivores (the Centetidae) and fertilization begins within the Graafian follicle.

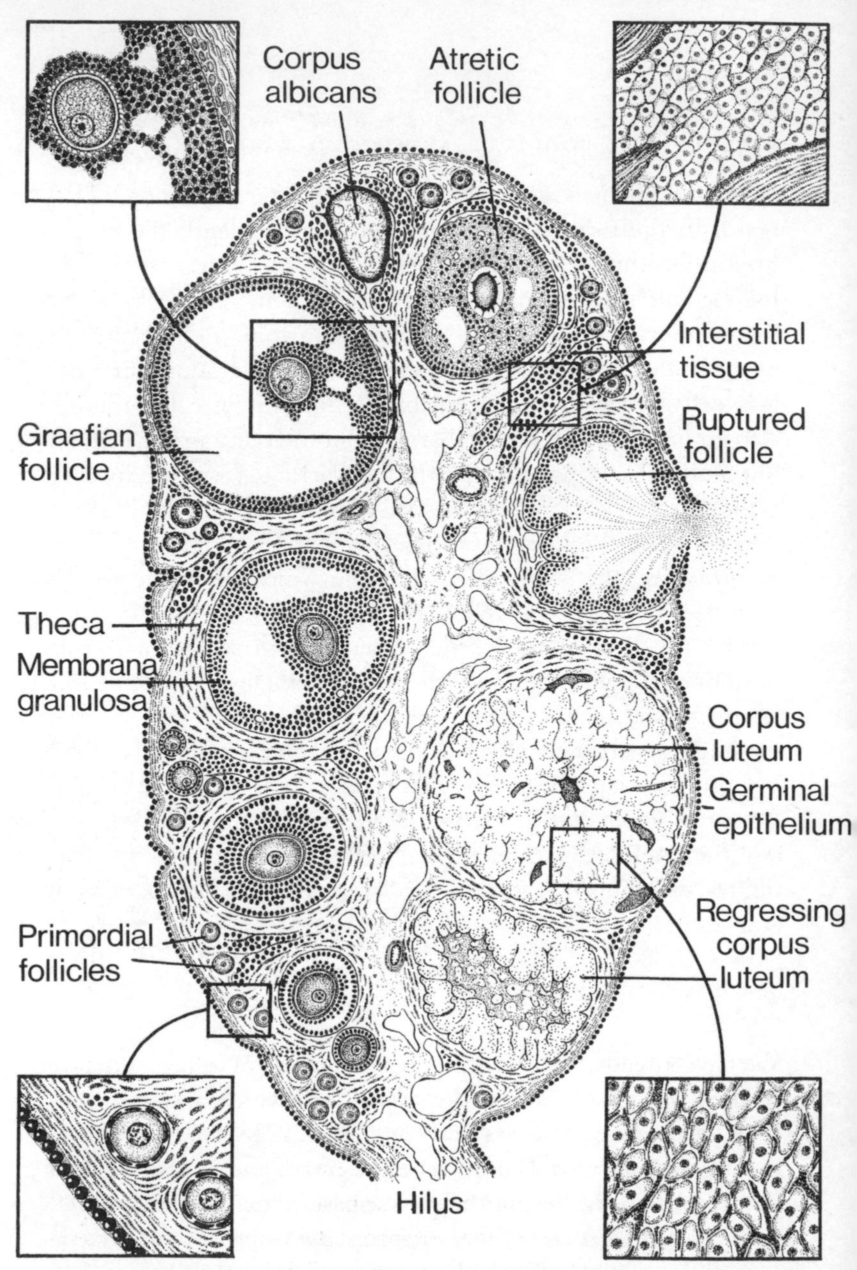

Fig. 2-12. Idealized drawing of the structure of the mammalian ovary showing follicles at various stages in their development and the formation and regression of corpora lutea. (Fig. 12-7 from C. D. Turner (1966) *General Endocrinology*, 5th ed. p. 400, W. B. Saunders Company.)

FATE OF THE FOLLICLE AFTER OVULATION

Dramatic changes occur in the follicle after ovulation. They result in the formation of a true endocrine gland, the corpus luteum, within the cortex of the ovary (see Fig. 2-12). The follicle initially collapses to a fraction of its former size and the membrana granulosa is thrown into folds. The remnants of the antrum are rapidly obliterated by proliferation of granulosa cells which become transformed into lutein cells; there is also an infiltration with capillaries from the theca interna. Some of the thecal cells pass in with the capillaries to form subdividing walls, the trabeculae, but the main component of the gland is derived from the granulosa. The lutein cells contain secretory granules and the yellow pigment from which the name corpus luteum is derived (i.e. 'yellow body').

The function of the corpus luteum is to secrete a steroid hormone, progesterone, which is important in controlling the length of the reproductive cycle in many species and also for maintaining pregnancy (see Book 3, Chapter 4). The gland has a finite life depending on whether or not pregnancy ensues. Towards the end of its life the secretion of progesterone ceases and the lutein cells degenerate. Regression is completed by the infiltration of fibroblasts which convert the gland into a mass of scar tissue (corpus albicans; Fig. 2-12).

ATRESIA

We have already remarked that few oocytes and follicles survive to the stage of ovulation, the great majority being eliminated by a degenerative process known as atresia. By far the greatest 'wastage' of germ cells occurs before birth, and affects oogonia at interphase and during the course of mitotic division, as well as oocytes at all stages of meiotic prophase (especially pachytene and diplotene). In the human ovary, for example, some five million oogonia and oocytes are eliminated from the ovaries between the 5th month of gestation and full term. The factors responsible for such waves of atresia are obscure, although genetic defects and errors in metabolism are probably involved.

That genetic defects are important in the induction of atresia is shown from studies of mutant mice. Mutations at the *W* locus appear to induce degeneration of germ cells shortly after they have colonized the genital ridges. Similarly, human females devoid of one X-chromosome per cell (XO or Turner's syndrome) are characterized by the absence of oocytes after birth. The ovaries of fetuses with this genetic constitution are initially normal but the germ cells are eliminated during the last couple of months of prenatal life. Turner's syndrome in women is thus accompanied by total and permanent sterility. The condition is less deleterious in female mice in which XO germ cells not only survive but can give rise to viable offspring.

Circumstantial evidence for a genetic influence on prenatal atresia derives from the observation that morphological changes affect the chromosomes long before cytoplasmic defects can be detected. But if genetic defects were the only cause of atresia, this would imply that the load of mutations and chromosomal aberrations would be enormous; therefore it is unlikely to be true. Other errors in the germ cells, attributable to defects in metabolism or shortage of nutrients (due to inadequate follicular cells or paucity of blood supply) are probably important contributing factors.

Atresia continues to deplete the population of oocytes in the ovaries of postnatal mammals. The population declines rapidly during prepubertal life and then at a lower rate throughout adult life; these points were discussed early in this chapter. The rate of decline in the population of oocytes varies considerably between species and even between strains within a single species. In CBA mice, for example, the number of oocytes reaches zero when the animals are about 300–400 days old, whereas mice of the RIII and A strains retain some 1000 oocytes in their ovaries at the same age. Significantly, hybrids of the CBA strain and A stock have a higher population of follicles during middle to old age than either of the parent groups.

The earliest signs of atresia in 'growing' and Graafian follicles are: (*a*) condensation of chromosomes and wrinkling of nuclear envelope in the oocyte; and (*b*) pyknosis of the nuclei of granulosa cells, which are then detached from the membrana granulosa and become free-floating in the follicular fluid. The oocyte may appear to complete its meiotic divisions abnormally ('pseudomaturation') or undergoes fragmentation so that it superficially resembles a morula. Pseudomaturation changes are occasionally observed in atretic primordial follicles, the granulosa cells of which usually survive while the oocyte is phagocytosed by macrophages. There can be no doubt that the rate of atresia is controlled by gonadotrophic hormones. Thus following ovariectomy the level of circulating FSH in mice is seemingly unaffected and the remaining ovary undergoes compensatory hypertrophy so that as many 'growing' and Graafian follicles occur as in controls. The rate of ovulation for the ovary doubles and the same number of offspring per litter is recorded. But the reproductive life-span and *total* number of offspring born to mice are halved since no 'new' oocytes can be formed to replace those in the extirpated ovary (these features are discussed more fully in Book 4, Chapter 5).

The rate of decline in the oocyte population with increasing age is considerably reduced, although not altogether prevented, by hypophysectomy. Transplantation into intact mice of the ovaries from those that were hypophysectomized 300 days previously, leads to a resumption of follicular growth and oocyte depletion by atresia. Injections of stilboestrol into hypophysectomized immature rats retards atresia affecting growing follicles whereas treatment with exogenous gonadotrophins (PMSG) hastens follicular atresia. The effect of PMSG varies with increasing dose from a follicle-stimulating effect, to atresia, and then to luteinization.

Exposure of the ovaries to ionizing radiation has the opposite effect to hypophysectomy; the rate at which oocytes are eliminated is initially increased. But the process of atresia following irradiation is not necessarily the same as that occurring spontaneously in oocytes. Radiation induces chromosomal breaks which may or may not undergo repair. The damaged cells are eliminated within hours or days of treatment, whereas those that are unaffected or have undergone repair may persist for prolonged periods, to be ovulated and fertilized. The dose of radiation required to destroy a given population of oocytes varies considerably between species and also depends upon age. Primordial oocytes in rats and mice are highly sensitive, while those in guinea pigs, Rhesus monkeys and possibly the human female, are fairly resistant to X-irradiation. The results of a recent study indicate that the control of radiation-induced atresia by hormones may be similar to that occurring spontaneously. We have at present no idea of the nature of atresia, nor how long is needed for an affected cell to be eliminated from the ovary. Grossly abnormal cells may undergo lysis *in situ* ('pools' of degeneration), or be phagocytosed by surrounding somatic cells. Clearly the process of atresia is regulated by gonadotrophins and probably steroids as well, although the precise controlling mechanisms are obscure. One of the most fundamental problems in reproductive biology that still remains completely unsolved is why one oocyte should be selected to undergo meiosis and ovulation, while its immediate neighbours suffer atresia and are eliminated from the ovary. Furthermore, why should the granulosa envelope of one particular oocyte be stimulated to undergo mitosis and growth when other apparently identical cells are unaffected? These forms of 'germinal selection' do not seem to operate in the male and cannot be due to genetic faults alone since the implied mutation frequency would be enormous. New techniques need to be devised if further studies of the processes of atresia and germinal selection are to enhance our knowledge of such fundamental issues.

A great deal of interest among biologists is stirred by the growth and differentiation of the oocyte. This is understandable, for not only is it the largest of the body's cells but it also has the potential to develop into a complete new individual. It may even achieve this feat unaided by the spermatozoon – we have no good evidence that mammals cannot sometimes arise parthenogenetically – and so must carry within itself all the necessary machinery. Fittingly enough, oogenesis includes meiosis, a complex process that bestows uniqueness on the individual.

SUGGESTED FURTHER READING

A quantitative and cytological study of germ cells in human ovaries. T. G. Baker. *Proceedings of the Royal Society of London*. B **158**, 417–33 (1963).

Electron microscopy of the primary and secondary oocyte. T. G. Baker. In *Advances in the Biosciences* 6. Schering Symposium on Intrinsic and Extrinsic Factors in Early Mammalian Development, Venice, 1970. Ed. G. Raspé. Vieweg-Pergamon (1971).

The fine structure of oogonia and oocytes in human ovaries. T. G. Baker and L. L. Franchi. *Journal of Cell Science* **2**, 213–24 (1967).

The uptake of tritiated uridine and phenylalanine by the ovaries of rats and monkeys. T. G. Baker, H. M. Beaumont and L. L. Franchi. *Journal of Cell Science* **4**, 655–75 (1969).

Mammalian eggs in the laboratory. R. G. Edwards. *Scientific American* **215**, 72–81 (1966).

Marshall's Physiology of Reproduction. Ed. A. S. Parkes, (especially chapter by Brambell), 3rd edition. London; Longmans (1956).

Reproductive Physiology of Vertebrates. A. van Tienhoven. Philadelphia; W. B. Saunders Co. (1968).

The Chromosomes. M. J. D. White. London; Methuen (1952).

Sex and Internal Secretions. Ed. W. C. Young. 3rd edition, 2 vols. Baillière, Tindall and Cox (1961).

The Ovary. S. Zuckerman, (especially chapter by Ingram), 2 vols. London; Academic Press (1962).

3 Spermatogenesis and the spermatozoa

V. Monesi

Male germ cells or spermatozoa are produced in the testis by the process known as spermatogenesis. They are first formed at puberty, but they represent the culmination of events that begin early in embryonic life. We will take up the story of germ-cell development in the male from the moment of gonadal differentiation in the embryo, and study its progress up to the establishment of spermatogenesis in adult life. The events following the release of spermatozoa from the testis – including sperm life in the epididymis, transport of spermatozoa through the male and female genital tracts, and ejaculation, together with information on the function of accessory organs and the chemistry of the semen – are set out in Chapter 5. The effects of hormones on spermatogenesis are discussed in Book 3.

THE TESTIS BEFORE AND AT PUBERTY

In the first chapter of this book, we saw how the cellular ancestors of the sex cells, called primordial germ cells or gonocytes, originate from the yolk sac entoderm (in the 21-day-old embryo in man) near the developing allantois, then migrate by active amoeboid movements along the dorsal mesentery and reach their final location in the genital ridges some time before sexual differentiation. During migration, the primordial germ cells multiply rapidly, and they then become associated with cellular elements arising from the growth of the coelomic epithelium (wrongly called germinal epithelium) to form cord-like structures, the primitive 'sex cords', which are embedded in the mesenchyme tissue of the developing gonad. In man, the transformation of the indifferent gonad into a testis or an ovary begins near the end of the 2nd month of development (18 to

25 mm embryo). If the embryo is to develop as a male, the primitive sex cords continue to elongate and to penetrate into the underlying mesenchyme; they then join together, and lose their contact with the surface epithelium. This medullary position of the sex cords enables one to distinguish at this stage the male gonad from the female gonad, in which the germ cells occupy the cortex of the developing organ. During fetal life the primitive testicular cords, as they are now called, are composed of proliferating primordial germ cells scattered among smaller 'supporting cells', which are the precursors of the Sertoli cells. The primordial germ cells are large elements with a basophilic cytoplasm and a spherical homogeneous nucleus containing one or two large nucleoli. They are always isolated from the basement membrane by the more numerous supporting cells, which are cuboidal or columnar in shape, and form a multilayered tissue attached to the basement membrane. The supporting cells are smaller than the primordial germ cells and have a deeply stainable nucleus with coarse masses of chromatin, surrounded by a thin rim of cytoplasm.

In the connective tissue between testicular cords there appear the interstitial cells. They originate from fibroblast-like cells, and histochemical evidence shows that they contain enzymes capable of taking part in steroid metabolism. In man these cells appear as early as the 12th week of fetal life, and multiply rapidly until birth. Surprisingly, they diminish swiftly during the early postnatal period, and it is only some time around puberty that increased numbers of the cells, precursors of the future Leydig cells, appear in the testis. The temporal correlation between the function of Leydig cells (production of steroid hormones) and the beginning of spermatogenesis at puberty is still rather unclear.

During early postnatal life and the growth period, the testis enlarges slowly, but there is little change in the histological differentiation of the primitive tubules until the time of puberty. During this period, the human testis is composed of solid cords about 50 μm in diameter containing two cell categories – the precursors of Sertoli cells and the cells of the germinal line. The latter consist of the primitive germ cells or gonocytes, which during the course of infancy and childhood undergo a series of changes leading to the development of early forms of a new kind of cell known as spermatogonia. Many primordial germ cells, however, degenerate during fetal and postnatal development without differentiating into spermatogonia. The primitive 'type A' spermatogonia appear in the human testis as early as 2 months after birth; they are smaller than the gonocytes, more irregular in shape and are attached to the basement membrane. At sexual maturity, the sex cords hollow out, thus becoming seminiferous tubules while the supporting cells stop dividing and differentiate into typical Sertoli cells. Soon spermatogenesis begins. The most striking event at the beginning of puberty is the proliferation of spermatogonia. All adult types of spermatogonia (identified as A_1, A_2, A_3, A_4, Intermediate, and B) appear consecutively at this time, followed by primary and secondary spermatocytes at the successive stages of meiosis, and eventually by spermatids which complete their development into spermatozoa at the end of puberty. The cyclic pattern of spermatogenesis is established immediately the process begins at puberty.

In the rat spermatogenesis begins on the 4th day after birth, with the transformation of gonocytes into type A spermatogonia ('stem cells'). These cells divide several times, then change into Intermediate and type B spermatogonia, which in turn produce primary spermatocytes. On the 15th day after birth the first generation of primary spermatocytes enters meiosis, while a second generation of stem cells initiates a new round of multiplication and differentiation. By 45–50 days after birth the first spermatozoa are produced. From these studies we infer that the cyclic changes of the seminiferous epithelium and the timing of the various steps of spermatogenesis, similar to those observed in the adult, are established very early after the beginning of spermatogenesis. In fact, in the immature rat, the duration of the cycle of the seminiferous epithelium is about 11 days, as in the adult (12 days), and the time for the production of the first spermatozoa from the stem cell is about 41–46 days, as in the adult (48 days).

Several problems remain to be solved. We still do not know what controls the differentiation of the indifferent gonad into a testis or an ovary. The relative roles of sex cells and somatic cells (intertubular cells or supporting cells) in this process remains to be elucidated in mammals. Also unknown are the mechanisms that delay meiosis until puberty in the male, and promote the start of spermatogenesis at the time of puberty. Some of these problems are discussed in Chapter 2 of this book, and in Book 2, Chapter 2.

THE SEMINIFEROUS EPITHELIUM

In the adult, the seminiferous epithelium is composed of two categories of cells: supporting or Sertoli cells, and germ cells at different stages of development (Fig. 3-1).

Sertoli cells

The Sertoli cells (Fig. 3-2) are the somatic elements of the seminiferous epithelium. They are large triangular cells with the base fixed to the basement membrane and the apex extending inwards, towards the lumen. The cytoplasm exhibits an elaborate system of thin processes surrounding all germ cells except the spermatogonial stem cells, which are in close contact with the basement membrane. The electron microscope has clearly shown that the Sertoli cells do not constitute a syncytium, as formerly believed, but that they are individual elements. At the boundary between adjacent Sertoli cells, and between Sertoli cells and germ cells, the apposed plasma membranes are clearly separated by a thin intervening space (Fig. 3-7). The nucleus is generally ovoid and appears relatively homogeneous, except for a large nucleolus flanked by two or three deeply stainable chromatin masses (chromocentres). The cytoplasm contains mitochondria, granular and agranular endoplasmic reticulum, and is rich in glycogen, glycoproteins and lipids which show cyclic variations. The cytoplasm of Sertoli cells in human testis contains slender spindle-shaped crystals called the crystalloids of Charcot–Böttcher, the chemical nature and function of which are unknown. Sertoli cells never divide in thn mature testis, and they are very resistant to X-rays and other ionizing radiations, and to various toxic agents that destroy the germ cells.

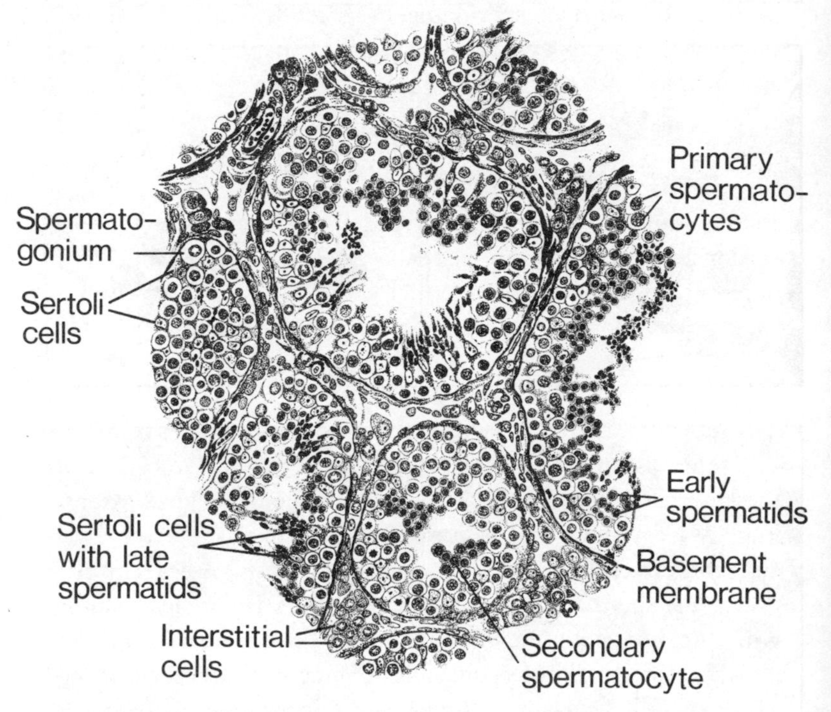

Fig. 3-1. Section of human testis to show the general morphology and the arrangement of germ cells at various stages of spermatogenesis in the seminiferous tubule. ×170. (Fig. 31-4 from W. Bloom and D. W. Fawcett, *Textbook of Histology*, 9th ed. p. 688 (1968) W. B. Saunders Company.)

The full function of Sertoli cells in spermatogenesis is unknown. They certainly provide mechanical support for the developing germ cells and play a role in the release of the

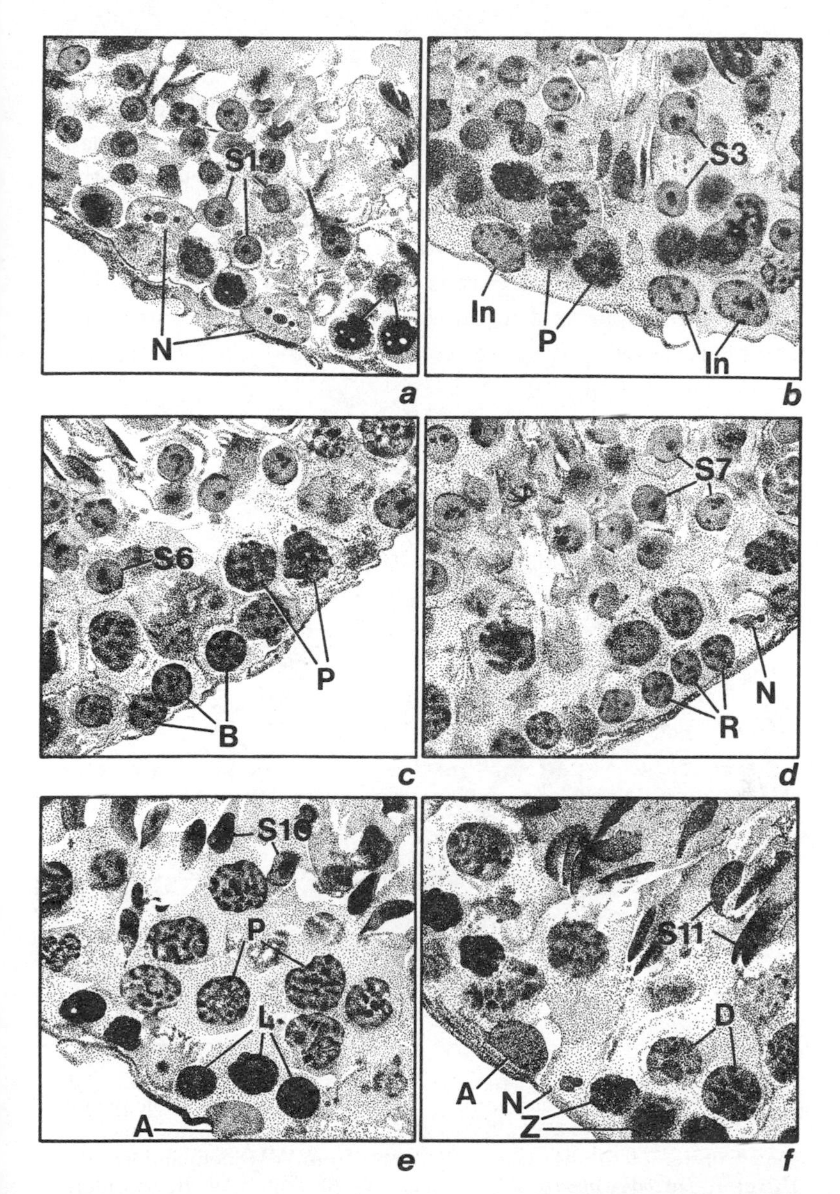

Fig. 3-2. Cross-sections of seminiferous tubules of the mouse stained by the PAS technique and haematoxylin. N: Sertoli cells; R. L, Z, P, D: primary spermatocytes at the pre-leptotene, leptotene, zygotene, pachytene, and diplotene stages respectively; S1, S3, S6, S7, S10, S11: spermatids at successive stages of development; A, In, B: type A, intermediate type and type B spermatogonia. (From V. Monesi. *J. Cell. Biol.* **22,** 521. Figs. 1–6 (1964).)

mature spermatozoa from the tubule and in the resorption of the 'residual bodies' which we will describe later. They may perhaps be concerned with the secretion or the transport of steroid hormones, or be an intermediate target system for the pituitary hormones. The close relationship between developing germ cells and the processes of Sertoli cells may mean that the Sertoli cytoplasm represents a medium through which inductive substances released by the germ cells diffuse and regulate the process of spermatogenesis.

Spermatogenesis lasts several weeks in all mammals, and may be subdivided into three phases: the first phase concerns the mitotic multiplication and maturation of spermatogonia, the second phase, meiosis, and the third phase, called spermiogenesis or spermateleosis, involves the transformation of spermatids into spermatozoa. The various cell stages of spermatogenesis are distributed from the periphery to the centre of the seminiferous tubule according to their age (Fig. 3-1).

Spermatogonia

Several types of spermatogonia can be identified in testicular sections (Figs. 3-2 and 3-14). The 'dust-like' or type A spermatogonium has an ovoid nucleus containing very fine chromatin granules distributed along the nuclear envelope, and a centrally located nucleolus. In the human testis there is a second class of type A spermatogonia, the 'dark type', which is characterized by a deeply stained dust-like chromatin and a pale-staining nuclear vacuole. In other mammals, such as the rodents, the bull, the ram and the boar, some spermatogonia exhibit morphological features that are intermediate between those of types A and B cells. These Intermediate type spermatogonia are characterized by fine chromatin granules attached to the nuclear envelope.

These types of spermatogonia – the dark type A (A_1), the pale type A (A_2) and the type B spermatogonium in man, and the type A, the Intermediate and the type B spermatogonium in other mammals – represent successive stages in the development of spermatogonia. The type A spermatogonia give rise to other type A spermatogonia (A_1–A_4 have been identified in the mouse); of this progeny, some cells divide to form Intermediate and type B spermatogonia and eventually spermatozoa, while others remain undifferentiated to serve as stem cells for future cycles of germ-cell multiplication and differentiation.

This process, whereby during each cycle of spermatogonial multiplication new stem cells arise to replace those that differentiate into spermatocytes, is called 'renewal of stem cells'. The time during the multiplication cycle of type A spermatogonia, when these renewing stem cells arise, varies in different species (see Fig. 3-13 for the mouse).

At several points during the mitotic divisions of spermatogonia, cells degenerate, so that the total number of spermatozoa eventually produced is lower than expected from the number of type A spermatogonia at the start of spermatogenesis. Degeneration affects mainly the type A spermatogonia. Further degeneration occurs during the two meiotic divisions, so that the number of spermatids produced by each primary spermatocyte is lower than four. The loss in spermatid production is about 27–37 per cent.

Spermatocytes (meiosis)

The division of type B spermatogonia gives rise to primary spermatocytes which enter meiotic division. As in egg maturation (see Chapter 2) meiosis consists of two consecutive cell divisions accompanied by only one duplication of the chromosomes, so that the resulting cells (spermatids) contain half the somatic number of chromosomes. During the first division, the homologous paternal and maternal chromosomes, which are longitudinally split into two chromatids and paired side by side, separate into the two daughter cells, the secondary spermatocytes, which therefore contain a single (haploid) set of chromosomes. In the course of the second division, the two sister chromatids of each chromosome, which had been so far held together by the centromere, are separated into the daughter cells, the spermatids.

The primary spermatocytes resulting from the last spermatogonial division before meiosis, resemble type B spermatogonia except for a smaller size (Fig. 3-2). After a short G1 phase, these cells undergo DNA duplication, lasting about 14 hours in the mouse, which can be demonstrated by using the radioactive precursor ^{3}H-thymidine. This stage which precedes meiotic prophase is called 'pre-leptotene'. As meiosis progresses, the primary spermatocytes move away from the basement membrane to the central part of the seminiferous epithelium, and become larger. The nucleus of secondary spermatocytes is considerably smaller than that of diplotene primary spermatocytes and exhibits finely granular chromatin and two or three large masses of condensed chromatin.

In the male the two sex chromosomes are not homologous, and during zygotene probably pair end to end rather than side by side. The X–Y bivalent is clearly visible during meiotic prophase as a very condensed (heterochromatic) body closely applied to the nuclear envelope, and is embedded within the so called 'sex vesicle'. Separation of the sex chromosomes at the end of the first meiotic division gives rise to two categories of spermatozoa, those bearing the X-chromosome and those containing the Y-chromosome, so that the sex of progeny is determined at fertilization.

Electron microscopic studies have shown the existence of cytoplasmic bridges connecting groups of spermatogonia, developing spermatocytes and spermatids. These bridges arise from incomplete cytokinesis during the successive divisions of the spermatogonia and of the spermatocytes. Late in spermatid differentiation, the spermatids separate to give rise to individual spermatozoa.

DNA and RNA synthesis during meiotic prophase and the synaptonemal complex

We have seen that DNA replication in meiosis occurs during

the pre-leptotene stage. This pre-meiotic DNA synthesis reflects the replication of the chromosomes in two chromatids. Some further synthesis of DNA occurs during zygotene and pachytene, but its nature and function are unknown. One suggestion is that it may be concerned with the process of repair of chromosome breaks associated with crossing-over at pachytene, or with the mechanisms of chromosome pairing and formation of the synaptonemal complex at zygotene.

The synthesis of RNA during meiosis has been studied by autoradiography, with the radioactive precursor ^{3}H-uridine. These studies have shown that the heteropyknotic X–Y bivalent does not incorporate RNA precursors throughout meiotic prophase. This result implies that the sex chromosomes are inactive in RNA synthesis during male meiosis, which seems to be a characteristic feature of the condensed chromatin (heterochromatin) as compared to the dispersed chromatin. By contrast, the autosomes appear very active in RNA synthesis during meiotic prophase, the rate of synthesis being much higher during pachytene and early diplotene than during the preceding stages. This differential rate of synthesis may be correlated with the development of a 'lampbrush' organization of the chromosomes in late prophase, similar to that observed in amphibian oocytes during diplotene. The DNA loops in amphibian oocytes are known to be sites of intense genetic activity of the chromosomes during oogenesis (see Chapter 2). Some electron microscopic observations would in fact suggest that DNA lateral loops are formed around pachytene chromosomes during male meiosis also.

The nature and function of the RNA molecules synthesized during male meiosis are unknown; they remain associated with the chromosomes for an extraordinarily long time, and are rapidly released into the cytoplasm at diakinesis and prometaphase. This 'meiotic' RNA, released into the cytoplasm at the end of meiotic prophase, represents most of the RNA in the spermatid cytoplasm, and might direct the differentiation of late spermatids, which do not synthesize RNA.

The pairing of chromosomes at zygotene both in spermatogenesis and oogenesis is characterized by the appearance of a longitudinal structure located between the homologous chromosomes of each bivalent. This is the synaptonemal (or synaptinemal) complex. Each complex is composed of two dense lateral elements, parallel to each other, and an intermediate layer of lower density which is in turn subdivided longitudinally into two parts by a dense medial element. Each of the five parallel zones is about 200 Å thick. The synaptonemal complex forms part of the paired homologous chromosomes until diplotene and disappears when the homologous chromosomes disjoin (Fig. 3-3). It is interpreted as a specific structure involved in chromosome pairing and perhaps also in crossing-over.

Evidence suggests that the lateral elements of the synaptonemal complex originate from the joining in pairs of the single, dense axial cores of leptotene chromosomes. According to this view, the medial zone of the complex represents the pairing zone of the chromosomes. The lateral components are composed of microfibrils 20–200 Å thick which appear to be continuous with the adjacent microfibrils of the chromatin. The central element and the intermediate zones on each side of it are crossed by transverse filaments which interconnect the lateral elements at the same level (direct fibres) or at different levels (indirect or ladder-like fibres). The transverse arrangement of filaments in the central zone may represent chiasmata.

The chemical composition of the synaptonemal complex is still obscure. Whereas some cytochemical and enzymatic digestion experiments at the electron microscope level have revealed the presence of some DNA in the lateral elements, we still do not know whether the central zone separating the two main components also contains DNA. The importance of detecting DNA in this zone lies in the suggestive hypothesis that the zone represents the sites where genetic exchange takes place. The existence of filaments crossing the medial zone from one lateral element to the other might favour this theory. Treatment with DNA or protein inhibitors can prevent the

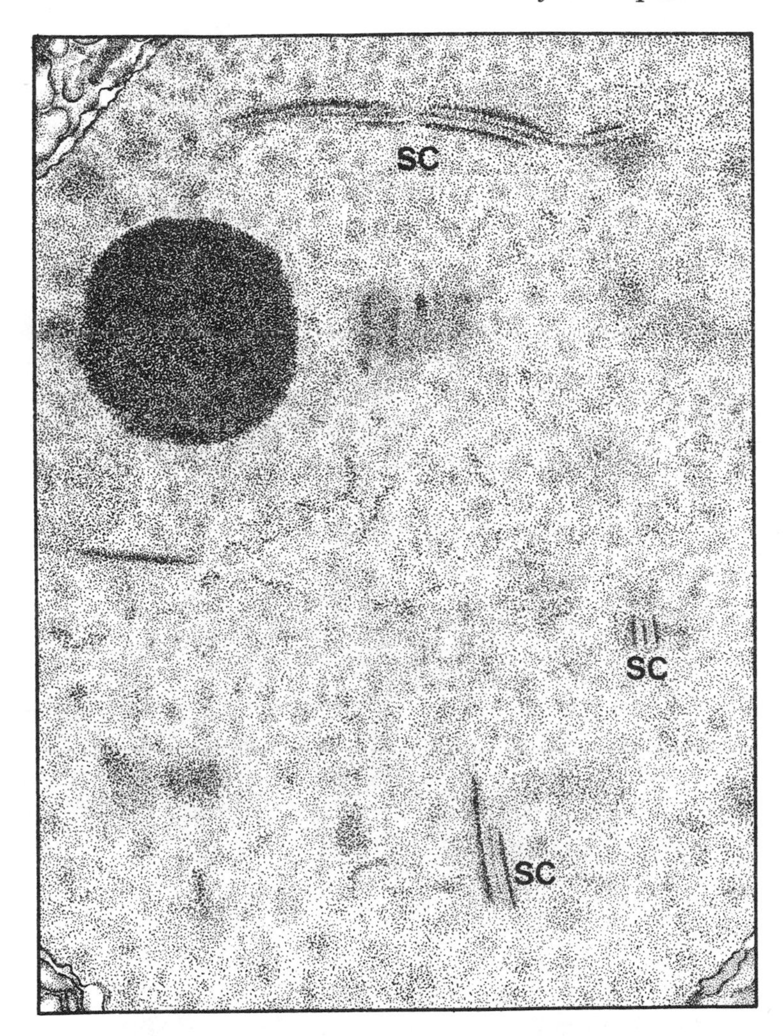

Fig. 3-3. Electron micrograph of a portion of a human spermatocyte nucleus showing three synaptonemal complexes (SC).

formation of the synaptonemal complex, and some DNA synthesis and protein synthesis has been shown to take place at zygotene, simultaneously with the formation of the synaptonemal complex.

Spermatids (spermiogenesis or spermateleosis)

The spermatids derived from the division of secondary spermatocytes undergo complex transformation into spermatozoa without cell division in a process called spermiogenesis or spermateleosis, during which specific structures, including the acrosome and the flagellum, differentiate.

The early spermatid has a small spherical nucleus 5 to 6 μm in diameter with a finely dispersed chromatin and several darker chromatin masses. The cytoplasm contains a Golgi apparatus near the nucleus and two centrioles at its periphery, as well as other organelles.

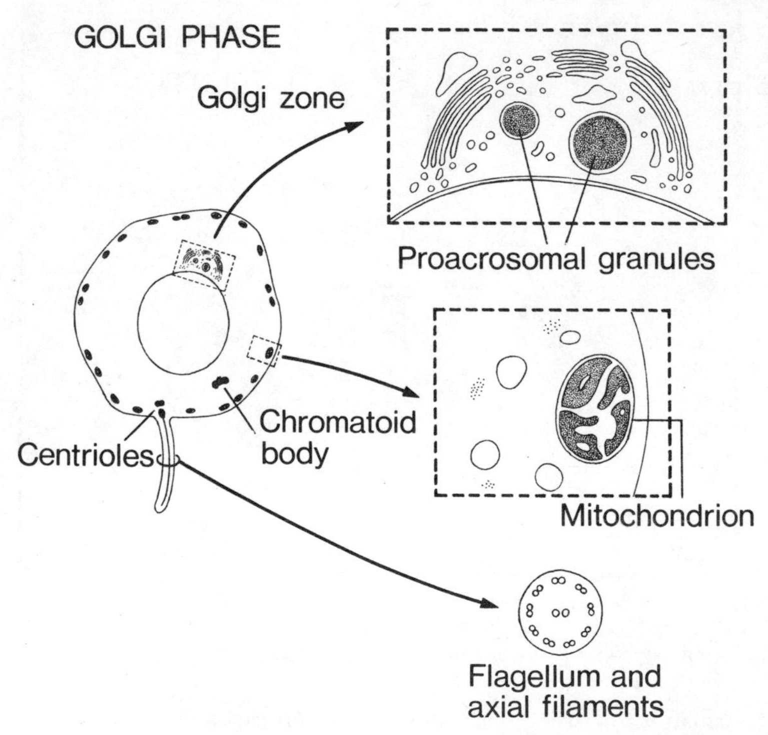

Fig. 3-4. (Above and opposite.) The Golgi phase and the cap phase of rat spermiogenesis. The submicroscopic structure is illustrated at the right in each figure. (From Y. Clermont. *Archs. Anat. micr. Morph. exp.* **56,** 7. Figs. 3 and 4. Paris; Masson et Cie (1967).)

Spermiogenesis can be subdivided into four phases, called the Golgi phase, the cap phase, the acrosomal phase and the maturation phase (Figs. 3-4 to 3-8). With light microscopy the morphology of spermatid development is well brought out by using the periodic acid–Schiff (PAS) reaction, which stains deeply red the glycoprotein material in the acrosomal structures. By this means, each phase of spermiogenesis may be subdivided into several stages. The total number of stages during the entire development of spermatids varies in different species; nineteen in the rat, sixteen in the mouse, fourteen in the monkey.

During the Golgi phase (Fig. 3-4) small granules called pro-acrosomal granules which are stained intensely by the PAS

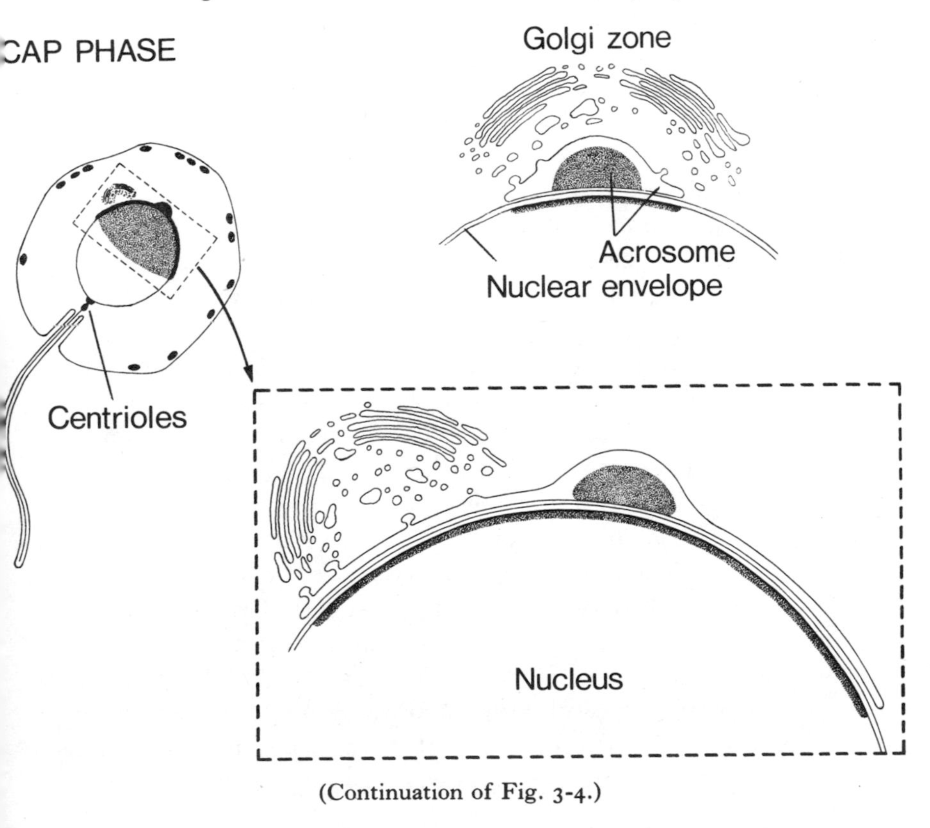

(Continuation of Fig. 3-4.)

reaction appear within the Golgi apparatus. Each granule is enclosed within a membrane-limited vesicle of the Golgi complex. As development progresses, the proacrosomal granules coalesce to form a single acrosomal granule, within its acrosomal vesicle, and later other vesicles from the Golgi apparatus contribute to the growth of the structure. The enlarged acrosomal vesicle becomes adherent to the nuclear envelope at the future anterior pole of the sperm nucleus (Fig. 3-7).

In the meantime one of the two centrioles, located at the opposite pole of the cell, gives rise to the flagellum which grows out of the cell though still enclosed within the plasma membrane (Fig. 3-4).

In the second or cap phase, the acrosomal vesicle spreads over the surface of the nucleus, forming a cap which covers the anterior half or two-thirds of the nucleus (Figs. 3-5 and 3-6). At this stage the acrosomal apparatus appears to consist of two zones surrounded by a membrane derived from the Golgi region.

The two centrioles migrate from the periphery of the cytoplasm inward toward the nucleus. One, the proximal centriole, becomes localized at the posterior pole of the nucleus, opposite the acrosome. The other, the distal centriole, acts as a basal corpuscle for the flagellum which has in the meantime elongated (Fig. 3-4). Until this time the flagellum consists only of the axial filament complex which is composed of two separate tubules in the centre and nine peripheral pairs of tubules arranged circularly around them (Fig. 3-4).

The acrosomal phase (Figs. 3-5 and 3-7) is characterized by profound modifications of the acrosome, the nucleus and the flagellum. The nucleus moves from the centre to the periphery of the cell, becomes elongated and slightly flattened, and its chromatin condenses progressively into coarse dense granules. The spermatid now rotates so that the acrosome becomes directed toward the wall of the seminiferous tubule. The acrosomal apparatus adapts to the shape of the nucleus, and so varies in shape and size in different species according to the form taken

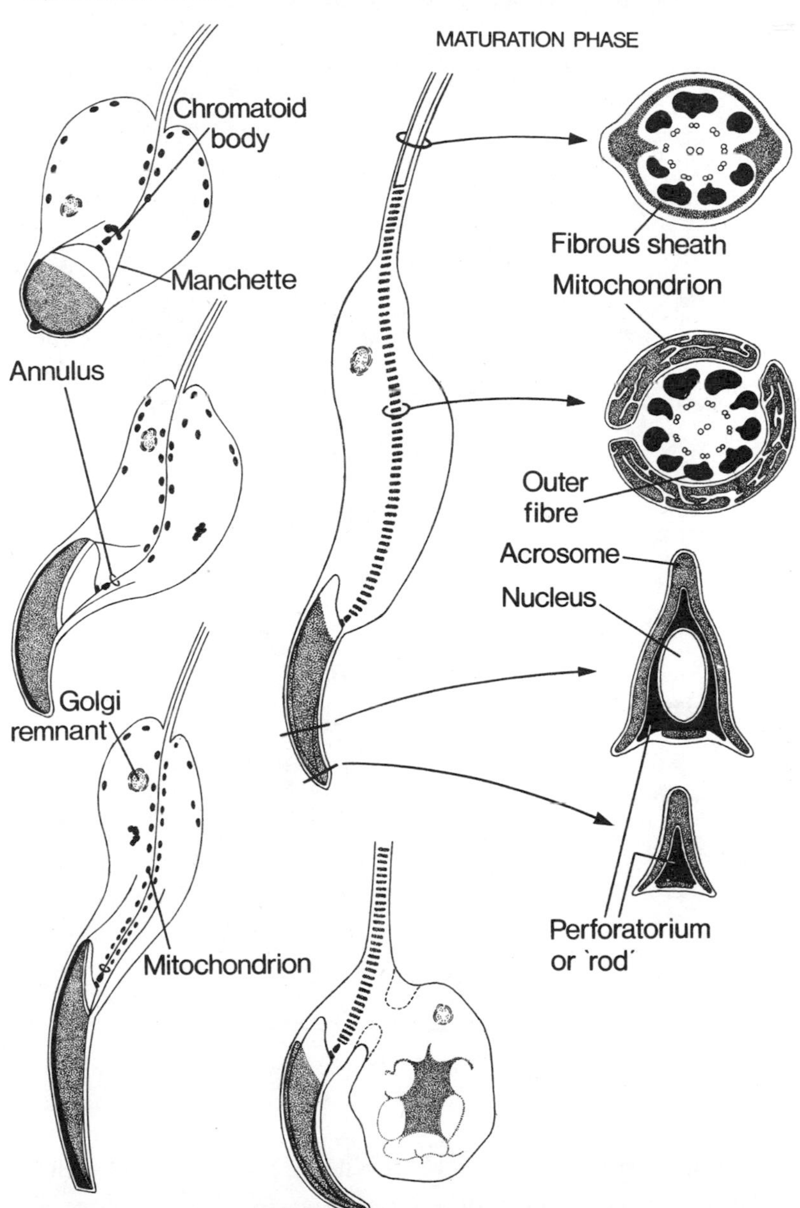

Fig. 3-5. The acrosomal phase and the maturation phase in the development of rat spermatids. On the extreme right, the submicroscopic structure of some cell organelles is shown. (From Y. Clermont. *Archs. Anat. micr. Morph. exp.* **56,** 7. Figs. 5 and 6. Paris; Masson et Cie (1967).)

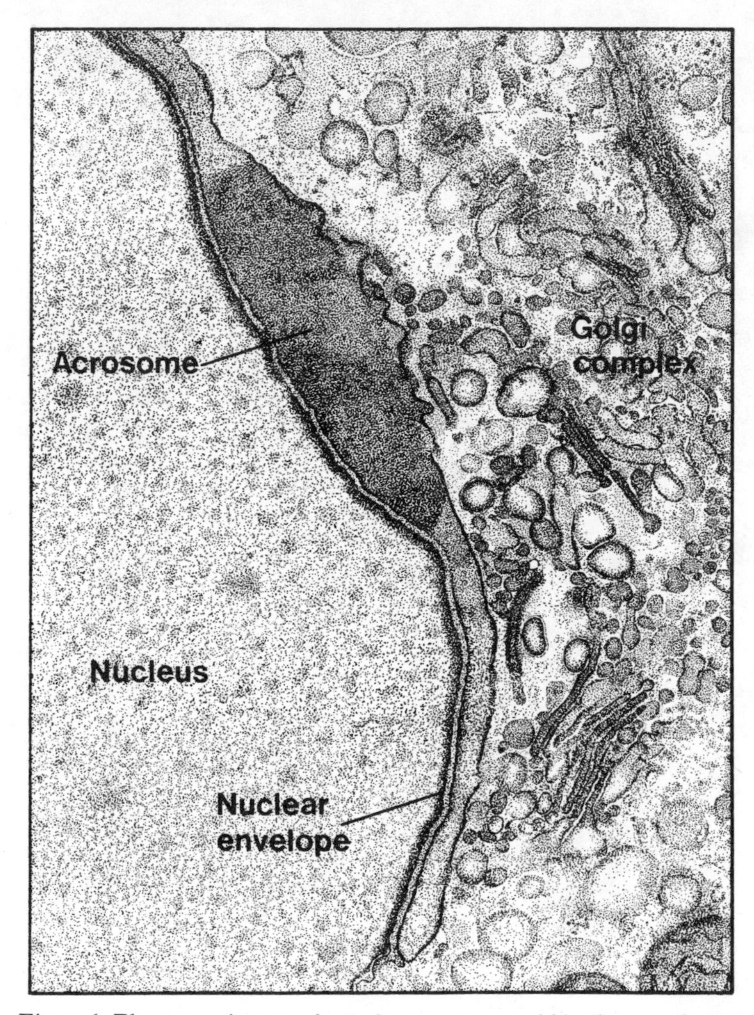

Fig. 3-6. Electron micrograph of a human spermatid in the cap phase of spermiogenesis. (Original by courtesy of L. Zamboni and M. Stefanini.)

by the elongating nucleus; it also becomes condensed in appearance.

As development progresses, the bulk of the cytoplasm is displaced behind the caudal pole of the nucleus and elongates to

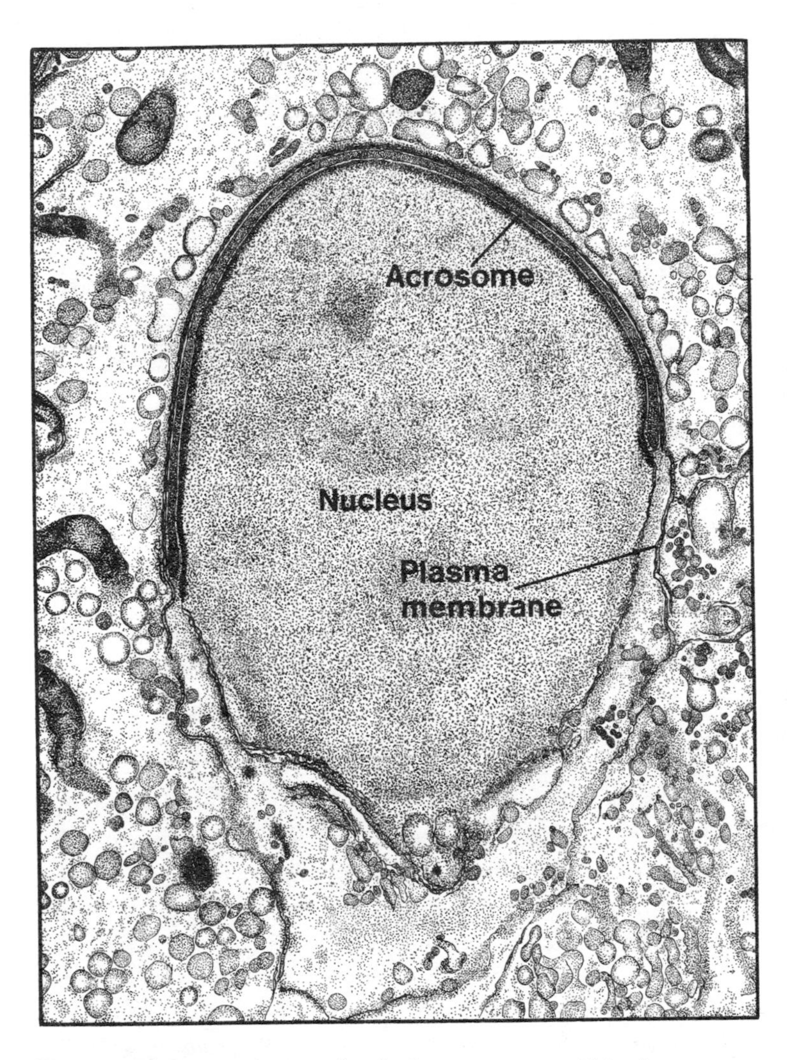

Fig. 3-7. Electron micrograph of a human spermatid in the acrosome phase of spermiogenesis. The spermatid is enclosed by a cytoplasmic process of a Sertoli cell. The apposed plasma membranes of the spermatid and the Sertoli cell are clearly visible. (Original by courtesy of L. Zamboni and M. Stefanini.)

surround the proximal part of the flagellum. In the cytoplasm, a special structure, the chromatoid body, which is present also in the preceding stages, approaches the distal centriole forming a ring-like structure called the annulus (previously termed the 'ring centriole') around the flagellum. The chromatoid body stains intensely with iron haematoxylin; with the electron microscope it appears to be composed of moderately dense filamentous or granular material. The mitochondria, which so far have been distributed at the periphery of the cytoplasm, migrate toward the portion of the flagellum inside the cytoplasm and surround it (Fig. 3-5). They will develop later into the mitochondrial sheath of the sperm middle piece. At about the same time, cytoplasmic microtubules associate to form a cylindrical structure, called the caudal sheath or manchette, which appears to be inserted over the nucleus and encircles the initial part of the flagellum (Fig. 3-5).

During the maturation phase (Figs. 3-5 and 3-8) the spermatid completes its transformation into a spermatozoon. The nucleus and acrosome take on the shape characteristic of the species. In the rat and mouse the nucleus has the form of a sickle. Its dorsal ridge is covered by the acrosome which is reduced in thickness over the rest of the nuclear surface, except for a narrow caudal–ventral zone (Fig. 3-5). In other species, like man, the nucleus takes on a flattened pyriform shape and the acrosome surrounds its anterior two-thirds (Fig. 3-10). The chromatin granules become coarser and denser until they coalesce into a compact homogeneous mass devoid of visible structure filling the entire nuclear area (Fig. 3-10). Nuclear condensation is preceded by complete arrest of the genetic activity of the chromosomes.

In some species, like the rat, a new structure, the 'rod', appears at this stage in the spermatid head. It is located on the apical part of the nucleus underneath the acrosome, probably in the sub-acrosomal space (see above) (Fig. 3-5). This structure seems to be lacking from the human spermatozoon. Its function and nature are largely unknown.

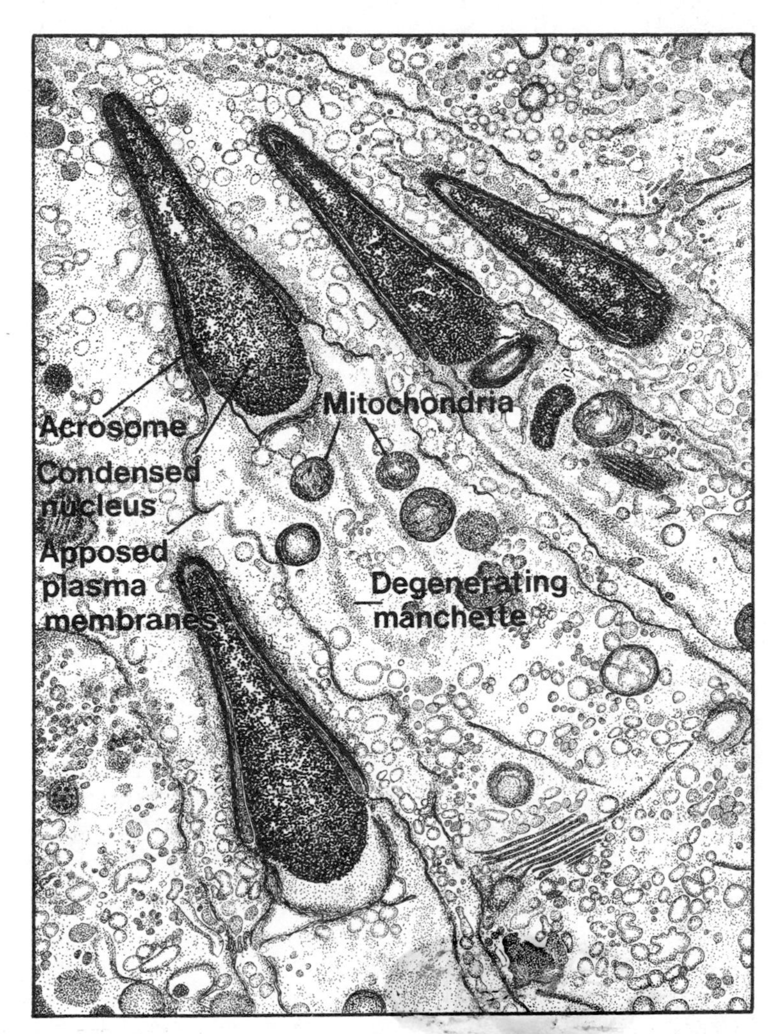

Fig. 3-8. Electron micrograph of human spermatids in the maturation phase of spermiogenesis. (Original by courtesy of L. Zamboni and M. Stefanini.)

The annulus, which surrounds the flagellum near the distal centriole, moves distally along the flagellum a certain distance from the distal centriole, and marks the limit between the middle piece and the principal piece of the flagellum. The manchette disappears during the final stage of spermatid development. At the beginning of this phase nine coarse fibres, or 'outer' fibres, are formed around the axial filaments; they run through the tail and terminate shortly before its end. At the level of the centrioles, the outer fibres fuse together forming a sheath covering the centriole, and may be continuous with the proximal centriole which could represent the origin for the flagellar beat. This sheath represents the point of attachment of the flagellum and constitutes the wall of the connecting piece of the neck (Fig. 3-5). At the level of the proximal part of the flagellum (the future middle piece of the tail) the mitochondria, which have a tubular shape, arrange themselves helically around the outer fibres, forming the mitochondrial sheath of the middle piece. In the distal part of the flagellum (the future principal part of the tail), a fibrous sheath is formed around the outer fibres. In the terminal portion of the flagellum, fibrous sheath and outer fibres are lacking (Fig. 3-5).

With the completion of spermiogenesis, most of the excess cytoplasm is cast off. The cytoplasmic mass, containing a great number of ribosomes, lipid droplets, degenerating mitochondria and Golgi membranes, constitutes the so-called residual body of Regaud; it is released in the lumen of the seminiferous tubule or is phagocytosed by the Sertoli cells (Fig. 3-5). A small portion of cytoplasm, the cytoplasmic droplet, remains, however, attached to the mature spermatid around the initial part of the flagellum; it will disappear during the final maturation of spermatozoa in the epididymis (Fig. 3-12). Soon after the shedding of the residual bodies, the sperm heads are released from the Sertoli cytoplasm, which has so far tightly embraced the developing spermatids, and the spermatozoa are set free in the lumen of the seminiferous tubule.

RNA synthesis in spermatids and gene action in gametes

Autoradiographic experiments with the radioactive RNA precursor ^{3}H-uridine have shown that the synthesis of ribonucleic acid ceases completely soon after the second meiotic division, during early spermiogenesis. The late spermatid and the mature spermatozoon are therefore genetically inactive. The genetic inactivation of the spermatid nucleus is accompanied by the arrest of synthesis of nuclear proteins and the elimination from the nucleus of the RNA molecules synthesized during meiosis, and of non-histone proteins. These molecular events coincide with the beginning of nuclear elongation and the onset of fibrillar transformation of the spermatid nucleus (step 8 or 9 of spermiogenesis in the mouse). At the end of spermiogenesis (step 11 to 14 in the mouse), when the structural reorganization of the spermatid nucleus is rapidly progressing, the 'meiotic' histone, which had been synthesized in pre-leptotene spermatocytes, is replaced by a new histone very rich in arginine, called 'sperm' histone. This is in turn probably substituted by yet newer histone types at the time of fertilization. 'Sperm' histone may be responsible for the condensation of the spermatid nucleus, or it may play a protective role by stabilizing the sperm genome against thermal denaturation or other alterations during 'storage' and passage of spermatozoa through the male and female genital tracts.

Cytoplasmic protein synthesis occurring in spermatids after the arrest of RNA synthesis is thought to be supported mainly by stable RNA molecules synthesized during meiosis and accumulated in the cytoplasm of the developing spermatids. This cytoplasmic RNA of 'meiotic' origin has been detected autoradiographically in the spermatid cytoplasm until late spermiogenesis. At the end of spermiogenesis, just before the release of spermatozoa from the Sertoli cytoplasm, the 'meiotic' RNA is eliminated from the cell within the 'residual bodies'. (Consequently the mature spermatozoon cannot synthesize protein – except perhaps within the mitochondria, but this is another story.) This evidence would indicate that the events of differentiation in spermiogenesis are mainly directed by gene products preformed during meiosis and utilized during the inactive stages until late spermiogenesis. If this is true, it follows that in a heterozygous animal both alleles must be expressed in the male gamete. However, there are suggestions of post-segregational (haploid) gene action in the mammalian gamete. These derive from two sources; (*a*) the demonstration of a paternal effect on the transmission of the *t* allele. The gene *T*, causing a tailless state, has a recessive allele, *t*, which seems to improve the fertilizing capacity of spermatozoa, for the segregation ratios, after mating with a heterozygous (*Tt*) male, are not Mendelian, more offspring than expected being born with the *t* allele. (*b*) The report that AB individuals (heterozygous for blood-group antigens) produce phenotypically A and B spermatozoa.

Apart from these rare cases, and other more controversial examples, post-segregational gene action does not usually seem to occur in mammals. It is important to note that male gametes may also be affected by gene products from the surrounding somatic cells (Sertoli cells) and the diploid germ cells (primary spermatocytes). The cytoplasmic bridges connecting groups of germ cells and the cytoplasmic processes of Sertoli cells enveloping the developing spermatids would favour this migration of informational molecules.

THE SPERMATOZOA

After leaving the seminiferous epithelium, the spermatozoa move slowly forward into the tubuli recti and the rete testis, and then through the efferent ducts to reach the epididymis. Here they can be stored for long periods, particularly in the tail of the epididymis.

Spermatozoa taken from the testis are not functionally mature. They acquire motility and fertilizing capacity during the passage through the epididymis, in a final process of maturation. Several structural and cytochemical changes of the spermatozoa occur in the epididymis: the acrosome modifies its shape slightly, the cytoplasmic droplet is discarded, the condensation of the nucleus increases, and the chromatin changes its chemical reactivity to various cytochemical reactions. Spermatozoa may undergo in some species a further physiological change during their transit of the female tract, which is called capacitation (see Chapter 5).

Morphology

The spermatozoon consists of a head and a tail. The tail may be subdivided into a neck, middle piece, principal piece and end piece (Figs. 3-9 and 3-11). Like other cells, the spermatozoon is enclosed within a plasma membrane.

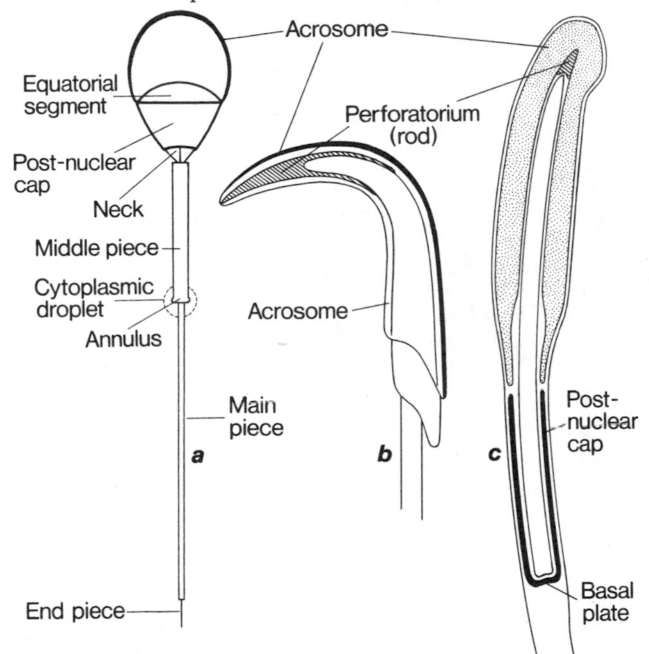

Fig. 3-9. *a*: General form of a spermatozoon – ungulate type. *b*: Rat sperm head showing acrosome and perforatorium or 'rod' (hatched). *c*: Vertical section of ungulate sperm head showing the relationship of the acrosome (stippled), perforatorium or 'rod' (hatched) and the post-nuclear cap (heavy line). The post-nuclear cap is here shown to be continuous with the basal plate, though they are in fact distinct structures. The thickness of the head relative to its length is exaggerated. (Fig. 1 of J. L. Hancock, Ch. 5 in *Advances in Reproductive Physiology*, vol. 1, Ed. Anne McLaren. London; Logos-Academic (1966).)

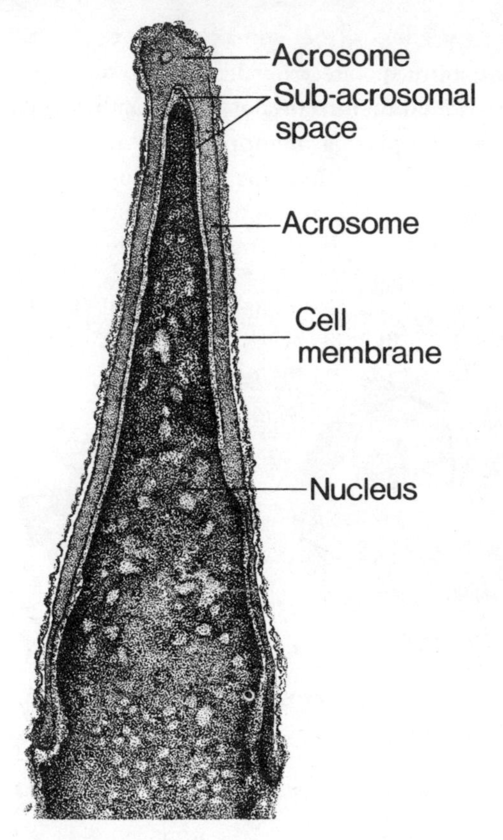

Fig. 3-10. Sagittal section of human sperm head from the ejaculate. (From M. Stefanini, C. De Martino and L. Zamboni. *Nature, Lond.* **216,** 173. Fig. 1*b* (1967).)

The shape of the sperm head varies greatly from species to species. In the human spermatozoon, the head has a flattened pyriform shape and measures 4 to 5 μm in length and 2.5 to 3.5 μm in width. The head is composed of two parts, the nucleus and the acrosome covering the anterior two-thirds of the nucleus (Figs. 3-9 and 3-10). In the rat and the mouse, the nucleus and acrosome form a sickle-shaped structure (Fig. 3-9). The nuclear chromatin appears extremely condensed, but the human sperm nucleus exhibits several areas of lower density or vacuoles without a limiting membrane (Fig. 3-10). The acrosome consists of an electron-dense matrix and is surrounded by the acrosomal membrane, of which an outer part is in close contact with the plasma membrane and an inner part is separated from the nuclear envelope by the so-called sub-acrosomal space. Posterior to the acrosome, the nucleus is covered by a laminar structure, showing often a periodic organization, called the post-nuclear cap, or better perhaps the post-acrosomal cap, the nature and function of which are unclear.

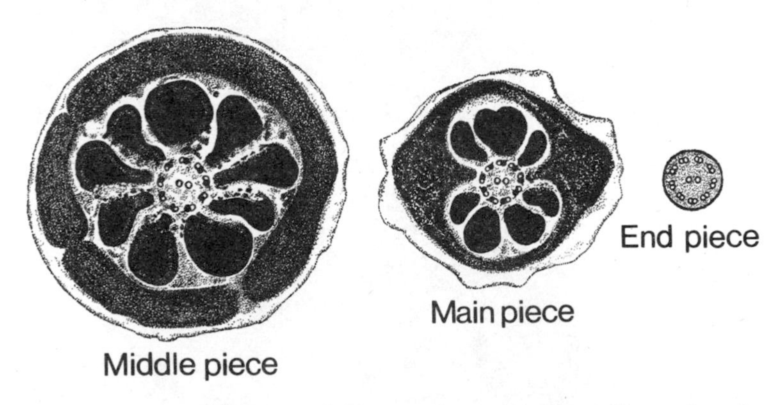

Fig. 3-11. Cross-sections of Chinese hamster sperm illustrating the structure of the three principal segments of the sperm tail. (From D. W. Fawcett. *Biol. Reprod.* **2,** Suppl. 2, 90. Fig. 8 (1970).)

The middle piece of the tail in the human spermatozoon is 5 to 7 μm long and about 1 μm thick, and extends from a slender connecting piece in the neck to the annulus. The principal piece is about 45 μm long and the end piece about 5 μm long. The thickness of the tail decreases gradually from the annulus to the end piece. The main components of the tail are the axial filament complex (axoneme), the coarse fibres, and the sheath, as already described (Figs. 3-11 and 3-12). In the rodents, two of the nine coarse fibres terminate in the principal piece, at a short distance from the annulus. In the middle piece the coarse fibres are surrounded by the mitochondrial sheath, which is arranged as a double helix, and in the principal piece

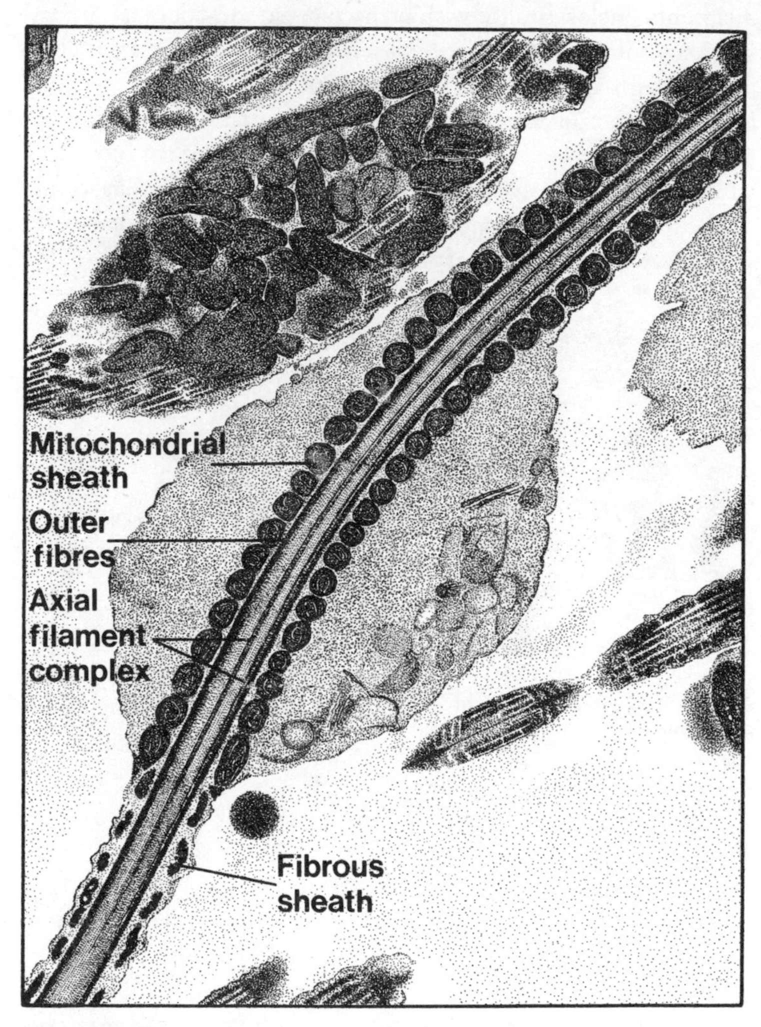

Fig. 3-12. Electron micrograph of a mouse spermatozoon taken from the cauda epididymis, showing the cytoplasmic droplet at the end of the middle piece, the mitochondrial sheath in the middle piece and the fibrous sheath in the main piece of the tail. (Original by courtesy of L. Zamboni and M. Stefanini.)

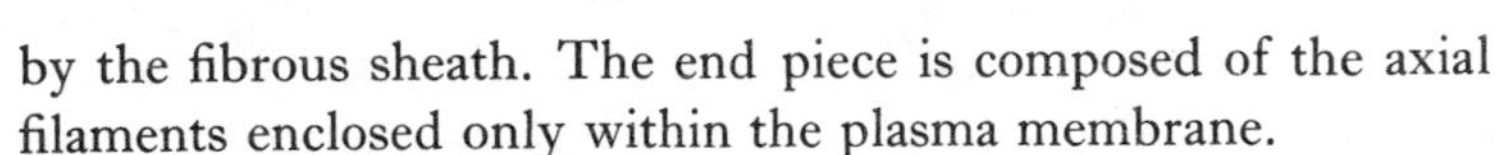

by the fibrous sheath. The end piece is composed of the axial filaments enclosed only within the plasma membrane.

Many authors have tried to separate X-containing from Y-containing spermatozoa by electrophoresis, counter-current centrifugation or density gradient sedimentation, with the aim of controlling the sex ratio, but no conclusive results have been obtained. X- and Y-bearing spermatozoa do not seem to differ significantly in shape and dimension. This is in line with the evidence that post-segregational (haploid) gene action does not occur, or is very rare in mammals. Difference in specific gravity and/or mass may exist between X and Y spermatozoa, owing to the different DNA content in the two sex chromosomes, but experiments involving insemination with spermatozoa separated into classes according to their sedimentation behaviour showed no effect on sex ratio. Recently the long arm of the Y-chromosome in somatic cells was found to exhibit an intense fluorescence with certain fluorescent compounds, such as quinacrine mustard and quinacrine hydrochloride. With the same technique, Y-bearing and X-bearing human spermatozoa (but not those of the bull, rabbit and mouse) can be readily distinguished by the presence in the former of a discrete fluorescing spot. The technique is not of immediate value for controlling the sex ratio, since spermatozoa must be fixed prior to staining.

Biochemistry and physiology

The sperm nucleus contains a haploid amount of DNA associated with a very arginine-rich histone in a ratio close to 1:1 by weight. Other acidic proteins (non-histone proteins), which are present in the chromatin of most cell types, are probably absent or very scarce in sperm chromatin. Mature spermatozoa contain only traces of RNA. There have been indications, based mainly on microspectrophotometric measurements, that deviations from the normal content of DNA may occur in spermatozoa from subfertile or senile individuals, but the data are still controversial.

The acrosome contains several carbohydrate materials, which are responsible for its intense staining by the PAS reaction and are associated with various enzymatic activities.

Chemical studies on detached acrosomes have detected a lipoglycoprotein, composed of a phospholipid and a glycoprotein. In the glycoprotein seventeen amino acids were demonstrated, as well as six main carbohydrate components: mannose, fucose, galactose, glucosamine, galactosamine, and sialic acid. The acrosome contains hyaluronidase, several lytic enzymes (including a trypsin-like enzyme) and acid hydrolases. Most of these enzymes are associated with the lipoglycoprotein complex which exhibits characteristic proteolytic and hyaluronidase activity. When added to newly ovulated rabbit eggs, the lipoglycoprotein complex prepared from ram, bull or rabbit spermatozoa, is capable of dispersing the cumulus oophorus and corona radiata, and sometimes dissolving the zona pellucida as well. The observations are consistent with the fact that the acrosome plays an essential role in the process of fertilization (see Chapter 5). The acrosome may be regarded as a modified lysosome, on the basis of its chemical composition and physiological function, and of its origin.

The sperm head contains other proteins, including a sulphur-rich keratin-like protein which probably constitutes the plasma membrane.

The axial filament complex is made up of several proteins, whose nature is not fully understood, though some are probably responsible for the contractility of the flagellum. The flagellum also contains adenosinetriphosphatase, an enzyme that cleaves adenosinetriphosphoric acid (ATP), thus liberating the energy required for motility, and acetylcholinesterase. Respiration and glycolysis provide the energy for the motor function of the spermatozoon. The application to spermatozoa of biochemical procedures designed for the extraction of myosin, actin and actomyosin from muscle have indicated the presence in the sperm flagellum of myosin-like protein ('spermosin'), actin-like protein ('flactin'), an actomyosin complex and even tropomyosin-like protein. Spermosin possesses high adenosinetriphosphatase activity, and undergoes a reversible interaction with muscle actin, as does the myosin from muscle. All this evidence strongly suggests a close similarity of flagellar proteins to the contractile proteins of muscle, but the functional and spatial relationship of these proteins to one another and to the structural units of the flagellum remain to be clarified.

The mitochondrial sheath is particularly rich in phospholipids (mainly lecithin and choline plasmalogen), bound to protein, and contains the complete cytochrome system and a number of oxidizing enzymes. A peculiar feature of the enzyme content of the germ cells is the presence of an additional isoenzyme of lactate dehydrogenase which is absent in somatic tissues. This isoenzyme is called LDH-X or LDH-IV because it is separated as band IV during electrophoretic separation, and may be under the control of a separate gene which is active only in primary spermatocytes.

The metabolic activity of spermatozoa and seminal plasma will not be discussed here, as this is a very large topic, save to mention that the two major types of metabolic processes in semen as ejaculated are respiration and glycolysis (fructolysis). These relate not only to the supply of energy for sperm motility, but also to sperm survival in the female reproductive tract.

SPERMATOGENIC CYCLES AND WAVES

Spermatogenesis begins with a stem cell or type A spermatogonium, which provides the starting point of a spermatogenic series. The kinetics of spermatogenesis are characterized by three remarkable phenomena, knowledge of which is essential for the comprehension of germ-cell development in the testis. These are (1) in any given segment of the seminiferous tubule the stem cells initiate the series of mitotic divisions preceding meiosis at extremely regular intervals (in the Sherman rat this interval is 12 days; in the mouse it is 8.6 days); (2) once the stem cells have initiated the process of multiplication and differentiation, each step of spermatogenesis has a fixed and constant duration; (3) in any given area of the seminiferous epithelium, before production has achieved the formation of spermatozoa, several new series of stem cells initiate at successive intervals new rounds of development which proceed concurrently. In the mouse, for instance, during the interval the stem cell takes to differentiate into a spermatozoon (that is, during the entire duration of spermatogenesis), four consecutive series of stem cells or A-type spermatogonia and their progeny develop concurrently, though spaced by the same time interval (see Fig. 3-13).

These three factors explain the typical arrangement of germ cells across the seminiferous epithelium. As mentioned earlier, in most mammals (except man – see below), the seminiferous epithelium is composed of concentric layers of germ cells that are at stages of development progressively more advanced from the periphery to the centre of the tubule. Each layer is composed of germ cells at exactly the same stage of development, and the various cell generations or stages of development are associated with one another in a well-defined and constant order. Each layer constitutes, therefore, a generation of germ cells which develops synchronously and in fixed correlation with the other generations. For instance, in the rat the mature spermatids (spermatids at step 19 of spermiogenesis) at the centre of the tubule are always found with primary spermatocytes at middle pachytene, with pre-leptotene spermatocytes, and with type A spermatogonia (Fig. 3-14).

In any given cross section of the seminiferous epithelium, the developmental stage of each cell generation and the combination of the different generations change with time as spermatogenesis proceeds. The time interval between two successive appearances of the same stage of development within each cell generation, or between two successive appearances of the same cellular association in any given area of the seminiferous tubule, is called the cycle of the seminiferous epithelium. From what was stated above, the duration of the cycle is a constant and corresponds to the interval between the cyclic beginnings of

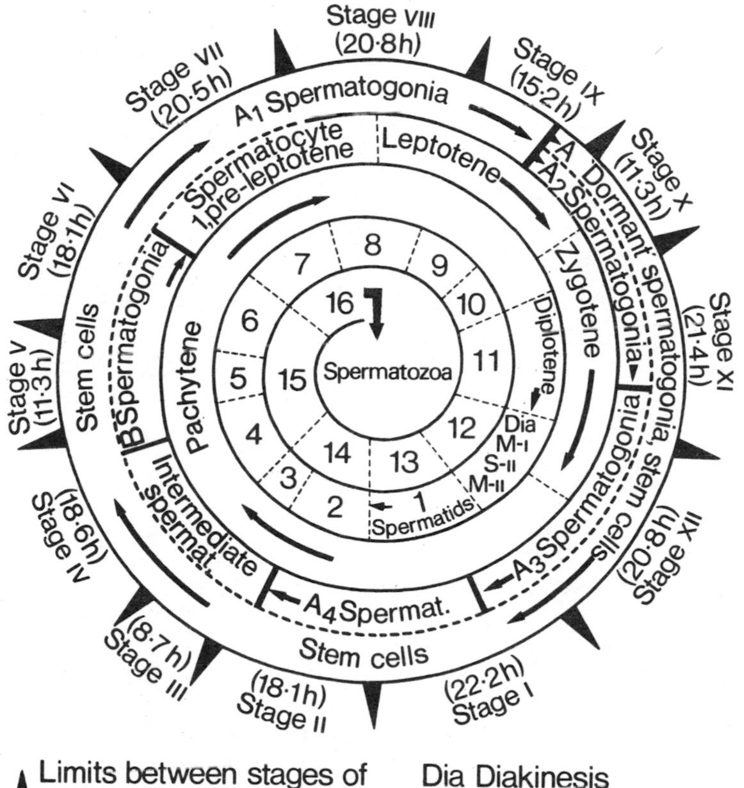

Fig. 3-13. Diagrammatic representation of the dynamics of spermatogenesis in the mouse. (Modified from E. C. Roosen-Runge. *Biol. Rev.* **37,** 343. Fig. 2 (1962).) Starting from the outer circle at 12 o'clock and proceeding clockwise, one can complete the entire development from stem cell to spermatozoon in four concentric turns. Each turn corresponds to one cycle of the seminiferous epithelium. The sectors limited by the spikes represent stages of the cycle, and show the cell composition of each stage, from the periphery to the centre of the tubule. 1 to 16 are the various stages of spermiogenesis as revealed by the PAS technique. Further description is given in the text. (From V. Monesi. *Arch. Anat. micr. Morph. exp.* **56,** 61 (1967).)

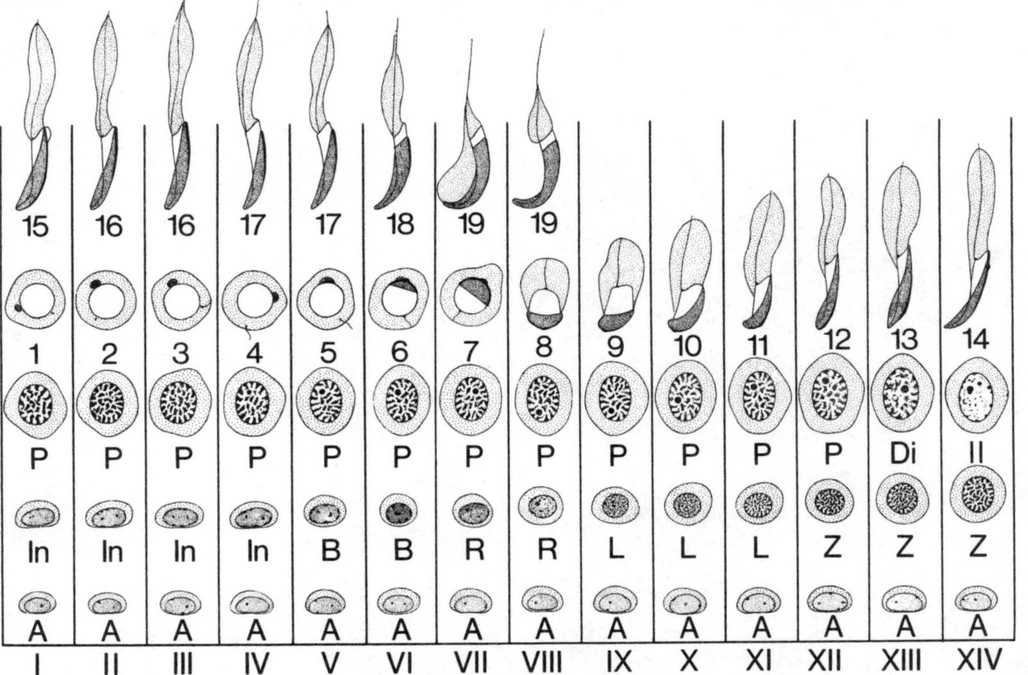

Fig. 3-14. Composition of the fourteen cellular associations observed in the seminiferous epithelium in the rat. Each column consists of the various cell types composing a cellular association or stage of the cycle (identified by roman numerals at the base of the figure). The various cellular associations succeed one another from stage 1 to stage 14. The succession of the 14 cellular associations constitutes the cycle of the seminiferous epithelium. *Abbreviations*: A, In, B: Type A, Intermediate and type B spermatogonia; R, L, Z, P, Di: primary spermatocytes at pre-leptotene, leptotene, zygotene, pachytene, and diakinesis respectively; II, secondary spermatocytes; 1–19, successive stages of spermiogenesis. (From B. Perey, Y. Clermont and C. P. Leblond. *Amer. J. Anat.* **108,** 49. Fig. 2 (1961).)

spermatogenesis from the stem cells, and to the interval between the stages of development of the various cell generations composing the concentric layers of the tubule. These concepts are illustrated diagrammatically in Fig. 3-13.

The arrangement of the germ cells in fixed combinations of cellular generations, homogeneous and synchronous in development, lends itself to a classification of the cycle into stages. Two principal methods of classification have been used. One is based on the meiotic divisions, the shape of the spermatid nucleus and the release of spermatozoa into the lumen of the seminiferous tubule. The other, which is more currently used, is based on the development of the acrosomal system. The second method depends on using the changes of the acrosomal system of the developing spermatids stained by the PAS reaction, as a key for the definition of successive stages of the cycle, which are commonly identified by roman numerals. By this method, the cycle of the seminiferous epithelium has been subdivided into several stages whose number varies in different species: fourteen in the rat, twelve in the mouse, thirteen in the golden hamster, twelve in the guinea pig and six in man. Each stage is composed of a fixed cellular association. The various cellular associations making up the cycle appear successively in any given area of the seminiferous tubule and this sequence repeats itself, cycle after cycle, indefinitely.

In most mammals, except man, the various cell associations or stages of the cycle occupy segments of the seminiferous tubule of different length. Segments of the tubule corresponding to successive developmental stages occur in an orderly sequence along the length of the seminiferous tubule. Each complete series of adjacent segments containing the typical cellular associations of the cycle is defined as a wave of the seminiferous epithelium (Fig. 3-15). Whereas the cycle concerns the changes occurring in time in any segment of the seminiferous epithelium, the concept of the wave refers to the arrangement of the cellular associations *in space* at any given moment along the length of the seminiferous tubule. In the rat, each seminiferous tubule contains 12 waves which measure 2.6 cm each on the average (with a range from 0.7 to 6 cm). In some places, the orderly sequence of the stages of the cycle along the tubule, called 'continuity of the segmental order', is disturbed by local reversal

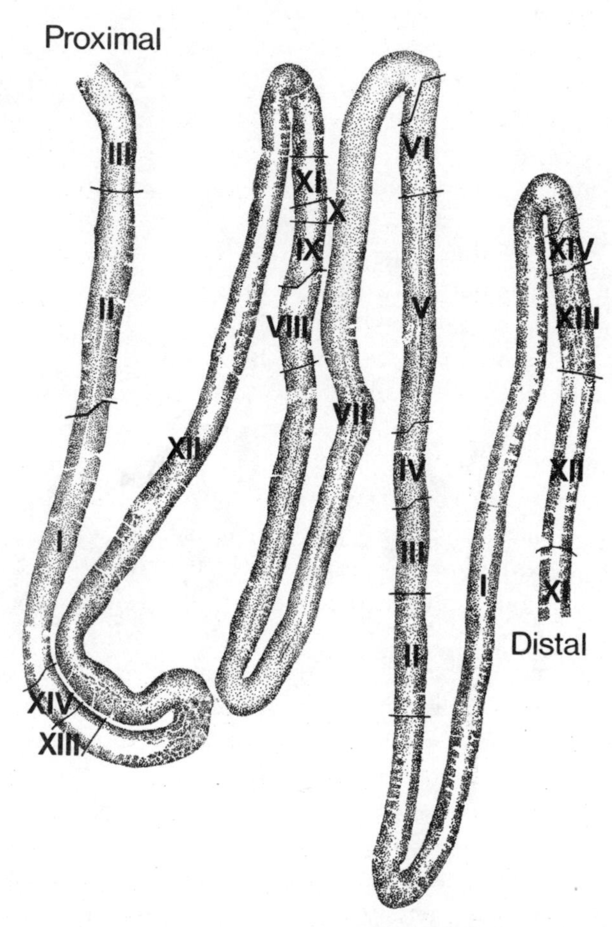

Fig. 3-15. Low power photograph of longitudinally cut seminiferous tubule of the rat, showing a wave without modulation of the seminiferous epithelium. ×25. The roman numerals indicate segments of the tubule showing the typical cellular association of the cycle of the seminiferous epithelium. (From B. Perey, Y. Clermont and C. P. Leblond. *Amer. J. Anat.* **108,** 49. Fig. 7 (1961).)

in the order of the consecutive stages. The irregularities are termed 'modulations' of the segmental order.

The cycle of the seminiferous epithelium in man

In contrast to the very regular ordering of the germ cells in most mammals, the disposition of germ cells in the seminiferous tubules of the human testis at first sight appears irregular. Because of this apparent disorder it was believed for a long time that in man there was no synchrony of germ-cell development nor a cyclic pattern of spermatogenesis, comparable to those found in most mammals. This has been found recently to be erroneous. By using a precise quantitative analysis, Yves Clermont has identified six well-defined cellular associations or stages, which were classified as stages of the cycle of the seminiferous epithelium. However, instead of taking up the entire cross-section of the seminiferous tubule, as in most mammals, these cellular associations occupy small and numerous areas of the epithelium, so that several different cell combinations or stages of the cycle may be seen in a single cross-section of a tubule.

As a result of the irregular distribution of the stages of the cycle along the seminiferous tubule, a 'wave of the seminiferous epithelium' comparable to that found in the rat and other mammals, does not exist in man.

TIME RELATIONS IN SPERMATOGENESIS

The duration of the cycle of the seminiferous epithelium and that of the entire process of spermatogenesis, from the stem cell to the release of spermatozoa into the lumen of the seminiferous tubule, have been determined for several mammalian species, including man. The method more currently used consists of labelling the DNA of spermatogonia and spermatocytes with the radioactive DNA precursor ^{3}H-thymidine and following the labelled cells by autoradiography as they differentiate into spermatozoa. The duration of the cycle and of spermatogenesis

TABLE 3-1. Duration of the cycle of the seminiferous epithelium and of spermatogenesis in various mammalian species.

Animal	Duration of the cycle (days)	Number of cycles during spermatogenesis	Duration of spermatogenesis (days)	Method employed
Rat (Sherman)	12 ± 0.2	4	48	^{3}H-thymidine
Rat (Sprague–Dawley)	12.29	4	51.6	^{3}H-thymidine
Rat (Wistar)	13.3	4	53.2	^{3}H-thymidine
Mouse	8.6 ± 0.2	4	34.5	Repopulation after X-irradiation
Rabbit	10.9 ± 0.1	4	43.6	^{3}H-thymidine
Ram	10.5	4.7	49	^{32}P
Monkey (*Macaca mulatta*)	—	6	—	—
Man	16 ± 1	4.6	74 ± 4	^{3}H-thymidine
Bull	13.5	—	—	^{3}H-thymidine

The time taken for the spermatozoa to pass through the ductus epididymis is about 8 days in the mouse, 11–14 days in the ram, 8–11 days in the bull, 14 days in the boar and 8–10 days in the rabbit.

varies with species (Table 3-1). One can also calculate the duration of the different stages of spermatogenesis. In the mouse, for instance, the spermatogonial mitoses take up about 8 days, meiosis about 13 days and spermiogenesis about 13 days (Fig. 3-13).

The duration of the cycle, i.e. the rate of germ-cell development, is a constant for any given species, and is not affected by changes of physiological conditions, such as elevation of temperature, or by hypophysectomy or X-irradiation. All these factors would influence the efficiency of spermatogenesis by inducing extensive cell degeneration without affecting the rate of residual spermatogenesis. The stage most sensitive to the effects of hypophysectomy is spermiogenesis, which is almost completely arrested during nuclear elongation and chromatin reorganization (step 7 of spermiogenesis in the rat).

The sex cells of the seminiferous epithelium are also very sensitive to infectious diseases, alcoholism, dietary deficiencies, local inflammation and X-irradiation. All these induce degenerative changes. The most sensitive to ionizing radiation is the spermatogonal population, followed by spermatocytes and spermatids. The Sertoli cells are, in contrast, remarkably resistant to ionizing radiation and to various noxious factors. Some type A spermatogonia also are resistant to very high doses of X-irradiation and to other toxic agents, and become thus able to produce a more-or-less complete regeneration of the seminiferous epithelium when the noxious factor is removed.

In man, the seminiferous tubules undergo gradual involution with advancing age. In the senile period, many seminiferous tubules display extensive atrophy. The atrophic tubules are depleted of spermatogenic cells and contain only Sertoli cells. However, in the remaining tubules spermatogenesis may continue without visible alterations until advanced age.

Spermatogenesis is a most intricate and involved process, but this is not surprising when one considers the remarkable degree of specialization of the spermatozoon. Its genetic material is only half that of somatic cells, but it has acquired many attributes that they do not possess. This nuclear weapon must travel great distances to reach its target, and in doing so it must penetrate several defence systems. Its high success rate underlies one of man's greatest problems.

SUGGESTED FURTHER READING

Cinétique de la spermatogenése chez les mammifères. Y. Clermont. *Archives d'anatomie microscopique et de morphologie expérimentale* **56**, 7 (1967).

A comparative view of sperm ultrastructure. D. W. Fawcett. *Biology of Reproduction* **2**, suppl. 2, 90 (1970).

Recent observations on the ultrastructure and development of the mammalian spermatozoon. D. W. Fawcett and D. M. Phillips. In *Comparative Spermatology*. New York; Academic Press (1970).

Spermatogenesis and the spermatozoa

Definition of the stages of the cycle of the seminiferous epithelium in the rat. C. P. Leblond and Y. Clermont. *Annals of the New York Academy of Sciences* **55**, 548 (1952).

Autoradiographic study of DNA synthesis and the cell cycle in spermatogonia and spermatocytes of mouse testis using tritiated thymidine. V. Monesi. *Journal of Cell Biology* **14**, 1 (1962).

Chromosome activities during meiosis and spermiogenesis. V. Monesi. *Journal of Reproduction and Fertility* Suppl. 13, 1 (1971).

The molecular architecture of synaptonemal complex. R. Wettstein and J. R. Sotelo. In *Advances in Cell and Molecular Biology*. The Maple Press Co. (1971).

The fine structure of monkey and human spermatozoa. L. Zamboni, R. Zemjanis and M. Stefanini. *Anatomical Record* **169**, 129 (1971).

Morphogenetic aspects of acrosome formation and reaction. J. C. Dan. In *Advances in Morphogenesis*, vol. 8. Ed. M. Abercrombie, J. Brachet and T. J. King. New York; Academic Press (1970).

Biochemistry of Semen and of the Male Reproductive Tract. T. Mann. London; Methuen (1964).

Sperm metabolism. T. Mann. In *Fertilization*, vol. 1. Ed. C. B. Metz and A. Monroy. New York; Academic Press (1967).

Advances in Reproductive Physiology, vol. 4. A. McLaren. London; Logos Press (1969).

Sperm motility. L. Nelson. In *Fertilization*, vol. 1. Ed. C. B. Metz and A. Monroy. New York; Academic Press (1967).

The Human Testis, vol. 10. Ed. E. Rosemberg and C. A. Paulsen. *Advances in Experimental Medicine and Biology*. New York; Plenum Press (1970).

4 Cycles and seasons
R. M. F. S. Sadleir

Human beings can give birth at any time, but most other mammals have a distinct mating or breeding season, and bring forth their young during a limited part of the year. The length of the breeding season, when copulation and conception occur, differs a great deal among wild mammals. Probably the shortest breeding season is that of the Arctic ground squirrel, which emerges from hibernation in late May and only copulates during the first two weeks of June. On the other hand, tropical mammals, such as the hippopotamus, can conceive in every month of the year, although some species can have a distinct breeding season even on the equator. In general, the time of mating is determined by the length of the pregnancy that follows and by the optimal time for giving birth, depending on weather and food supply. In temperate regions, most mammalian births begin in spring and continue into summer. Thus some small mammals such as hedgehogs and wood mice, with gestations of 30 and 25 days, respectively, mate and give birth during overlapping times of the year, whereas the birth season in larger species like the red deer occurs 8 months after the breeding season in October. Even human beings may show seasonal variation: in Europe there are more births from February to June than from August to December, while in the United States and South America this pattern tends to be reversed. The reasons for these trends are obscure.

Domestication has had the effect of extending the breeding season, and so some domestic mammals tend to breed continuously (cow, pig); others, however, retain elements of their original breeding seasons (dog, cat, horse), or may even remain seasonal breeders (goat, sheep). Commensal mammals – those dependent on man for food and shelter – may also breed continuously (rat, house mouse), and it is possible that man's agricultural activities have resulted in the continuous breeding of some wild species such as the rabbit.

PUBERTY

Breeding begins when animals reach puberty, the age when sperm formation starts and ovulations begin. The gonads of young animals can develop when pituitary hormones are administered, but before puberty the animal's own pituitary only produces very small amounts of gonad-stimulating hormones (gonadotrophins). At a certain age, usually associated with a specific body weight, gonadotrophin production of the pituitary increases, the ovaries and testes respond by releasing gonadal (sex) hormones, and these cause the external secondary sexual characteristics to develop. In boys, the voice deepens and facial and pubic hair appears; in girls (where this stage is called the menarche) menstrual cycles begin, breasts and pubic hair develop and body shape alters (see Fig. 4-1). These changes gradually take place over a period of 2–4 years, at different times in different people. Corresponding changes occur in other mammals, such as the development of antlers in deer and the mane in lions. In small rapidly growing species such as rodents, puberty appears within a few weeks of birth, so that the young animals may be able to breed almost at once. But in larger seasonal breeders such as deer, puberty may not be reached before the onset of winter, so the animals mature in the following year. Even larger mammals require from 3–7 years to reach puberty. A characteristic of our own species is the extremely long time it takes to achieve puberty, some 13 or 14 years, though the age of puberty has in fact been steadily declining over the last 100 years.

OESTROUS AND MENSTRUAL CYCLES

Once puberty has been reached there is much variation between species in the times when females will accept males and copulate, a state called *heat* or *oestrus*. The word 'oestrus' derives from the Greek *oistros*, a gadfly, and was used to describe the erratic and nervy behaviour of a cow when being attacked by such flies. A similarly nervous and touchy disposition characterizes many females when in heat, so that oestrus now refers to this stage of the reproductive cycle. Some species show alterations in the external genitalia, in addition to alterations in

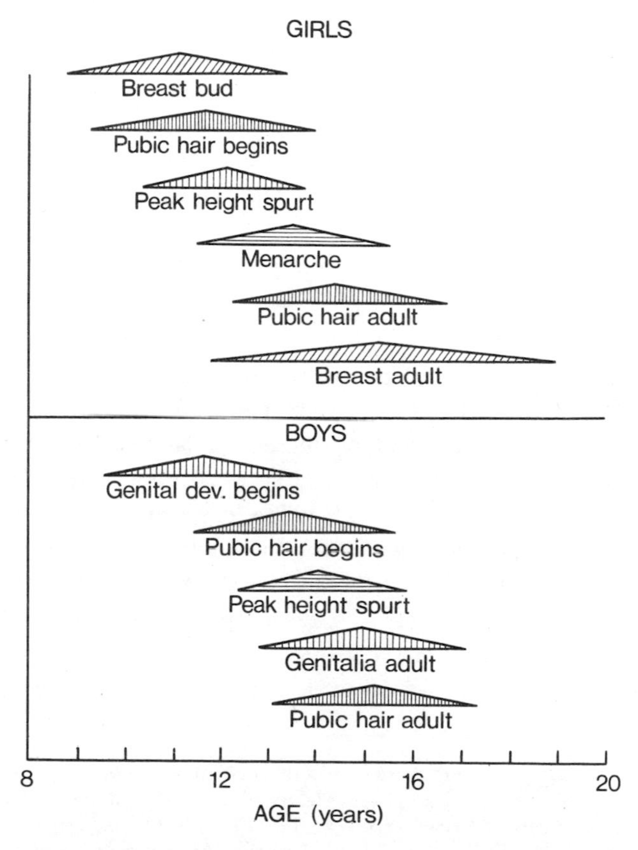

Fig. 4-1. The ages at which changes associated with puberty occur in girls and boys. (W. A. Marshall, *J. Biosoc. Sci.* Suppl. **2**, 31, Text-fig. 5 (1970).)

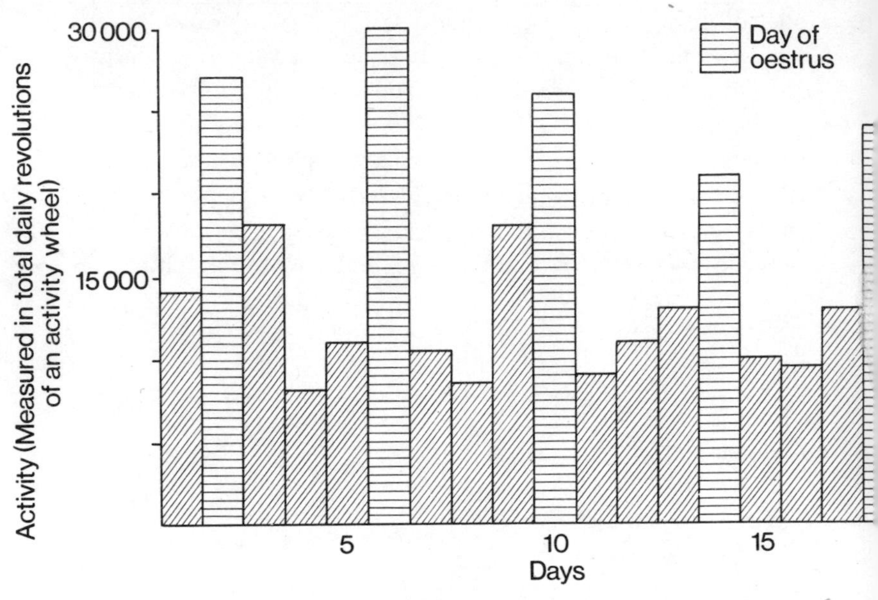

Fig. 4-2. Changes in activity levels during oestrous cycles in the rat.

temperament, but more commonly the oestrous female is detected only by her reaction to males and their attraction to her. There may be a general increase in activity of females in oestrus as shown for the rat in Fig. 4-2. In most mammals the oestrous period is short, with much longer intervening periods when the female will not mate; for example the oestrous cycle of a ewe is 16 days but she will only accept the ram for 24–36 hours during that period. The oestrous cycle as a whole is commonly divided into a sequence of phases – oestrus, metoestrus, dioestrus, pro-oestrus, oestrus – in each of which distinc-

---

Fig. 4-3. The different kinds of cells seen in vaginal smears taken at three stages of the oestrous cycle in the mouse. *Oestrus*: dead, flat 'cornified' cells almost exclusively. *Pro-oestrus*: mostly rounded epithelial cells with deeply stained nuclei. *Metoestrus*: large numbers of white blood cells, polymorphonuclear leucocytes. (During the fourth stage, dioestrus, a few cells of various kinds are found together with some stringy mucus.)

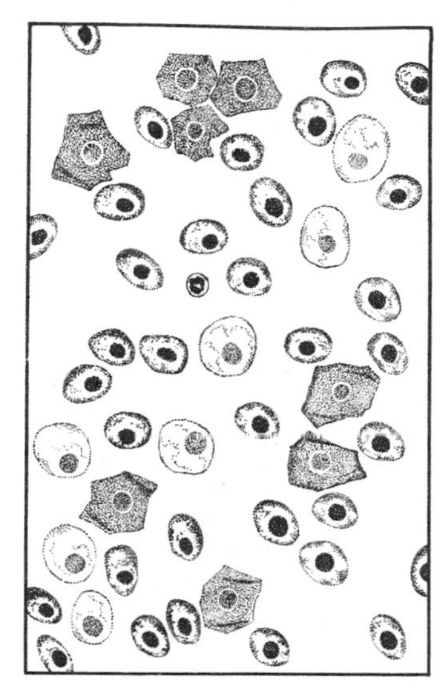

Pro-oestrus

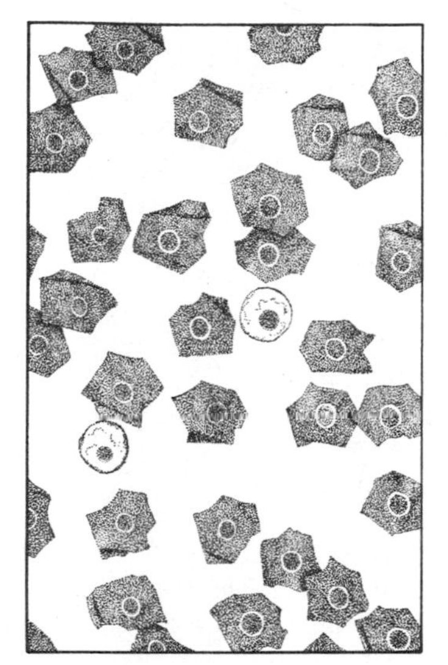

Oestrus

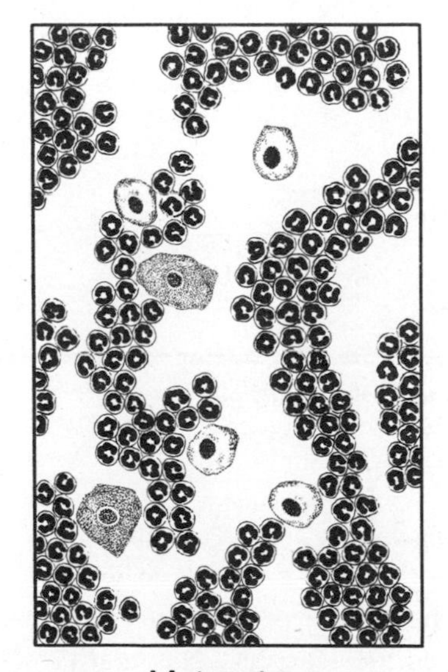

Metoestrus

tive changes are seen in the structure and function of the sex organs. In rodents, these phases are easily recognized by examining with a microscope cells scraped from the wall of the vagina (the 'vaginal smear') (see Fig. 4-3); cell changes are also seen in vaginal smears from other animals including man, but they are unfortunately not so distinctly categorized.

TABLE 4-1. Details of cycles in animals with spontaneous ovulation.

	Kind of cycle	Length of cycle	Length of oestrus
Opossum *Didelphis marsupialis*	Polyoestrous	28 days	1–2 days
Guinea pig *Cavia porcellus*	Polyoestrous	17 days	6–11 h
Mouse *Mus musculus*	Polyoestrous	4–5 days	3 h
Dog *Canis familiaris*	Monoestrous	—	9 days
Fox *Vulpes vulpes*	Monoestrous	—	2–4 days
Sheep *Ovis aries*	Polyoestrous	16 days	30–36 h
Cow *Bos taurus*	Polyoestrous	20 days	12–22 h
Pig *Sus scrofa*	Polyoestrous	21 days	2–3 days
Horse *Equus caballus*	Polyoestrous	22 days	4–6 days
Mouse lemur *Microcebus murinus*	Polyoestrous	45–55 days	2–5 days
Rhesus monkey *Macaca mulatta*	Menstrual	28 days	*
Chimpanzee *Pan satyrus*	Menstrual	35 days	*
Man *Homo sapiens*	Menstrual	28 days	**

* Copulation at almost any time, but receptivity greatest around the middle of the cycle.

** Copulation at any time.

Some mammals can potentially undergo a number of oestrous cycles during the breeding season, whereas other species may only come into heat once or twice a year – in effect, the latter actually lack breeding 'seasons'. *Seasonally polyoestrous* species, which include animals like the goat, ewe and mare, will thus have numerous, though clearly separated, opportunities for mating. Cats and other small mammals can even produce a sequence of litters in the breeding season. By contrast, *monoestrous* species, such as the fox and bear, can only produce one litter a year. Mammals such as the grey squirrel present a rather special case as they have two breeding and two birth seasons each year. Although thought to be monoestrous inside each season they are sometimes referred to as *dioestrous* species. The type of breeding pattern and the length of cycle of some representative species (wild and domestic) are shown in Table 4-1.

The oestrous period is usually of a specific length, but some species, such as the camel and the rabbit, exhibit a *continuous oestrus*. They remain prepared to accept the male for extended periods, and the act of copulation causes ovulation and terminates oestrus. Copulation either leads to pregnancy, or to the much shorter state of pseudopregnancy when fertilization or development fails (discussed later). Accordingly, in these animals oestrus is of very variable duration, and successive oestrous periods are separated by intervals that are also highly variable.

The special cycle of primates, the *menstrual cycle*, has the same hormonal basis but differs in that it terminates with menstruation – a destruction of the lining (endometrium) of the uterus which is associated with a loss of blood. The human menstrual cycle averages 28 days between menstruations. The cycles are more variable in length in teenagers and become very erratic after the late forties (Fig. 4-4). The smaller primates (monkeys, lemurs) have definite breeding seasons in the wild although they may breed at any time in captivity. The larger apes (chimpanzee, gorilla) breed continuously in nature and have no specific season of birth.

In seasonally breeding mammals, the males come into breeding condition prior to onset of oestrus in the female and may go out of breeding condition before the end of the female season, but most females will be pregnant by this time. Males will copulate at any time during the breeding season. In species in which the female breeding season lasts throughout the year, males are able to mate at any time.

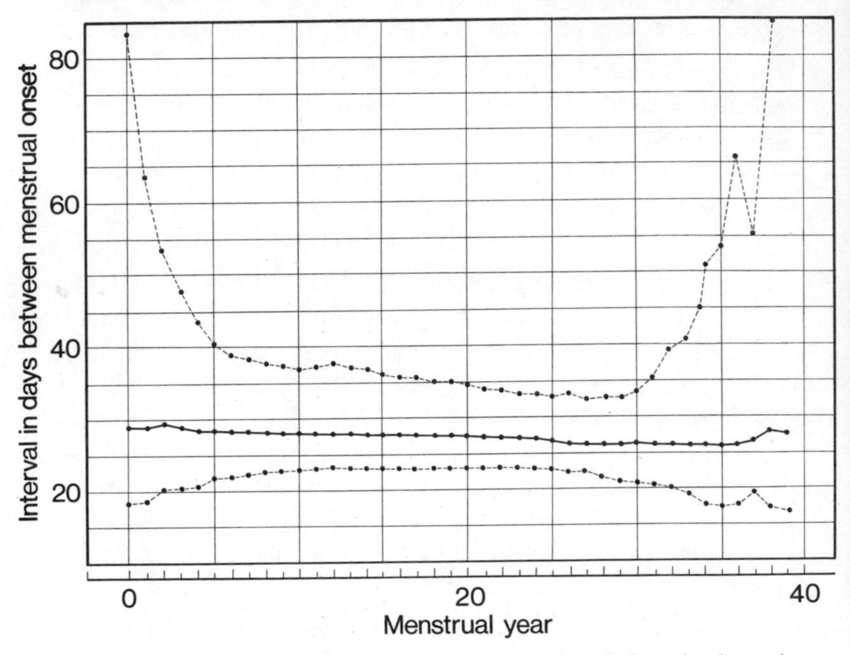

Fig. 4-4. The solid line shows median menstrual-cycle lengths throughout the reproductive life of women from menarche (year 0) to menopause (year 40). Ninety per cent of all cycles fall within the upper and lower broken lines. (From A. E. Treloar *et al. Internat. J. Fertil.* **12**, 77, Fig. 5 (1967).)

OVULATION

At one point during the oestrous or menstrual cycles an egg is released from the ovarian follicle (ovulation) and is transported to the top half of the oviduct, to a region identified as the ampulla. In non-primates this happens during (or sometimes just after) oestrus, and spermatozoa introduced into the vagina or uterus during copulation meet the egg in the ampulla, where fertilization takes place. The timing of the two events, oestrus and ovulation, is crucial because both egg and spermatozoon have very short lives. In most mammals so far investigated ovulation is *spontaneous* and occurs more or less at the same time in each oestrous cycle. For example in the laboratory rat the cycle is 4–5 days and oestrus takes about 14 hours; eggs are ovulated 8–11 hours after the start of oestrus. Ovulation in sheep is 18–24 hours after the beginning of oestrus, but in cattle about 10 hours after its end. The timing of ovulation is extremely variable in primates. In rhesus monkeys it generally takes place about 13 days after the onset of menstruation, but it can occur anywhere from 9 to 20 days after menstruation. Human beings are, if anything, even more variable: ovulation occurs generally some 14–15 days after the start of menstruation, but often at any time from the 9th to the 17th day. Variability in the time of human ovulation is mainly due to variability in length of the pre-ovulatory phase.

As we have already seen, other mammals ovulate in response to the stimulus of copulation – those with *induced ovulation*. In this curiously assorted group of animals, which includes the rabbit, ferret, cat, lynx, racoon, field vole, and camel, nervous stimuli arising from the act of copulation pass to the brain and this in turn brings about the discharge of the pituitary hormones that cause ovulation.

The detection of ovulation is of importance in artificial insemination (see Book 5, Chapter 1), in human contraceptive practice (see Book 5, Chapter 2), and in basic reproductive research. In mammals in which ovulation is spontaneous and fixed relative to the onset of oestrus, the time of ovulation can be approximately predicted from changes associated with oestrus, such as those occurring in the vaginal smear. This procedure is especially useful in laboratory animals such as the rat, mouse and golden hamster. In man, however, the vaginal smear does not show very clear-cut stages, and recourse must be had to other signs. Changes in parameters such as the glucose content of vaginal secretion, or the consistency of the cervical mucus, are often studied but these are rough and unreliable indicators of ovulation. Many women (and other female primates) show a slight but distinct rise in their basal body temperature just after ovulation (see Fig. 4-5); this does not

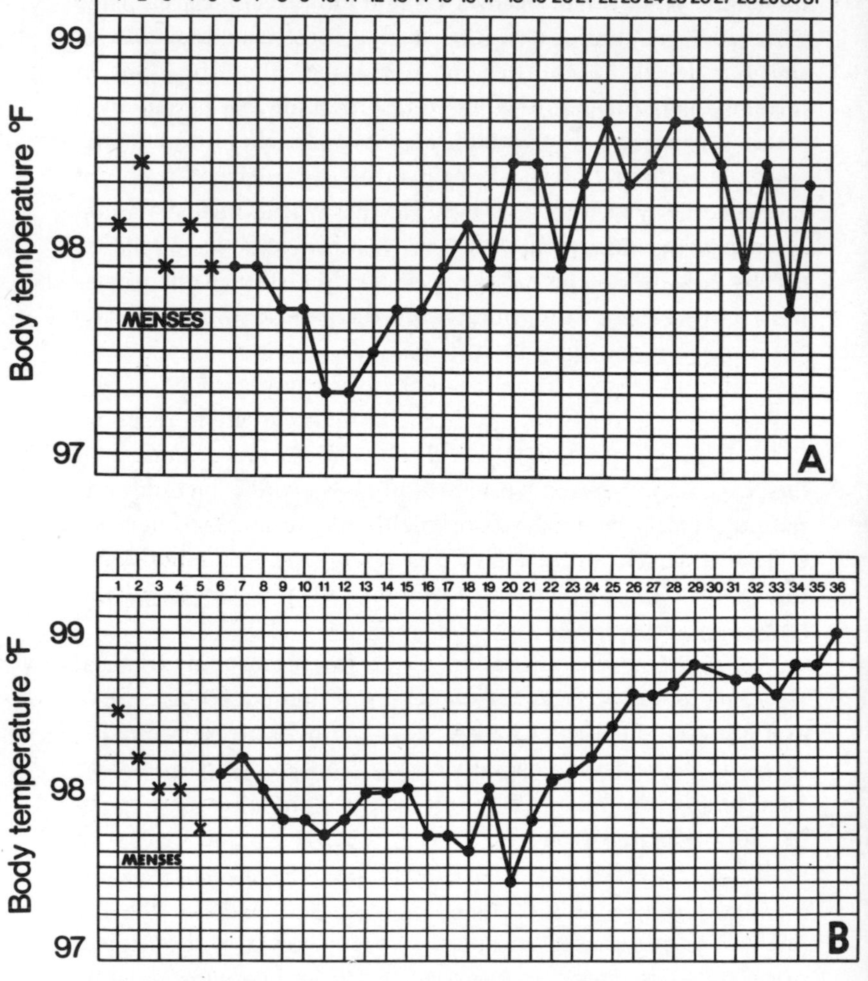

Fig. 4-5. Charts showing the changes in basal body temperature during the human menstrual cycle. Days are numbered from the start of menstruation. Chart A is of a cycle of the usual length, and Chart B a cycle that is unusually long. Ovulation is considered to occur at about the beginning of the continued increase in temperature. (Chart B from C. G. Hartman and J. H. Leathem, Fig. 31, Ch. 5 in *Mechanisms Concerned with Conception*, Ed. C. G. Hartman. London; Pergamon (1963).)

occur consistently in all women and is greatly affected by mild infections and other factors. Slight to severe back pains (*Mittelschmerz*) halfway through the cycle may indicate ovulation but not all women experience these. Daily estimations of hormone concentrations in the blood provide the best assessment, but require highly sophisticated methods. It should be emphasized that, despite considerable investigation into the problem, there is no completely reliable method for detecting externally the time of ovulation in man or other mammals. It can be done internally by direct rectal palpation of the ovaries in large species or by visual inspection of the ovaries with the aid of an instrument called a laparoscope.

PSEUDOPREGNANCY AND PREGNANCY

Oestrous cycles always cease during pregnancy, and the female tract is modified by hormones for the protection and nourishment of the developing embryo. In some mammals, such as the rat, mouse and dog, oestrous cycles will also cease during the condition of *pseudopregnancy* which can be invoked by artificial stimulation of the uterine cervix or by sterile mating. Although the egg has not been fertilized, the corpus luteum remains active and the uterus stays in a receptive state for about half the time of normal pregnancy, or even longer in the case of the bitch. Pseudopregnancy occurs in wild rabbit populations but rarely in other wild mammals.

We have already seen how the times of breeding and giving birth are related to the length of gestation. Species that are polyoestrous and have a short gestation can produce a sequence of litters during a breeding season. The frequency of litters is also governed by whether or not the species concerned can mate while suckling young (i.e. during lactation). Many mammals cannot conceive while nursing; there is no oestrus until the previous litter is weaned. Other species come into heat fairly soon after giving birth – they experience *post-partum oestrus*. These females can thus be simultaneously pregnant and lactating. The condition is very common in rodents and marsupials, and the best example amongst domestic species is the mare with her 'foal heat' about a week after foaling.

The length of *gestation* is roughly proportional to the size of the adult mammal. Table 4-2 shows that the larger animals have

TABLE 4-2. Some representative average gestation lengths in mammals.

	Days
Shrew (*Sorex araneus*)	20
Dormouse (*Muscardinus avellanarius*)	23
Rabbit (*Oryctolagus cuniculus*)	31
Cat (*Felis cattus*)	63
Dog (*Canis domesticus*)	63
Puma (*Felis concolor*)	93
Tiger (*Panthera tigris*)	105
Pig (*Sus scrofa*)	114
Sheep (*Ovis aries*)	150
White-tailed deer (*Odocoileus virginianus*)	204
Caribou (*Rangifer tarandus*)	230
Cow (*Bos taurus*)	280
Camel (*Camelus dromedarius*)	390
Black rhinoceros (*Diceros bicornis*)	540
Indian elephant (*Elephas maximus*)	623
Guinea pig (*Cavia porcellus*)	68
Chinchilla (*Chinchilla laniger*)	111
Coypu (Nutria) (*Myocastor coypus*)	130
Southern fin whale (*Balaenoptera physalus*)	365
Pacific grey whale (*Eschrichtius glaucus*)	365
Blue whale (*Sibbaldus musculus*)	365
Opossum (*Didelphis virginianus*)	13
Wallaby (*Protemnodon bicolor*)	35
Red kangaroo (*Megaleia rufa*)	33
Rhesus monkey (*Macaca mulatta*)	163 *
Baboon (*Papio conatus*)	187 *
Chimpanzee (*Pan satyrus*)	227 *
Man (*Homo sapiens*)	280 *

* Dated from onset of last menstruation.

the longer gestations, though these can vary a good deal in length. For example, the gestation period of the horse ranges from 325 to 350 days. Four groups of mammals do not fit the size rule. Rodents of the suborder Hystricomorpha, the guinea pig family from South America, have much longer gestations than other rodents of comparable body size. The huge whales have much shorter pregnancies than would be expected, and marsupials, as a part of their unique reproductive mechanisms, have a gestation period that is very short, although when added to the length of their 'extra-uterine gestation' in the pouch it comes approximately to what would be expected on the basis of their adult body size. Primates have rather longer gestation periods than other mammals of roughly equivalent size; human beings and cows have approximately the same gestation length!

Mammals with short gestations can produce several litters during a breeding season, but if gestation lasts more than a couple of months it is unlikely that the animal will be able to conceive again in the same year because the second litter would be born in late autumn or early winter. Thus the larger mammals will usually produce only one litter per year. Very large mammals with gestations of longer than one year (camel, rhinoceros, elephant) may only breed every second, third or fourth year, as they are in late pregnancy or early lactation when the breeding season comes around.

We remarked at the beginning that mammals with distinct breeding seasons have developed their time of copulation, parturition and weaning in relation to seasonal changes in their environment (and this point is taken up more fully in Book 4, Chapter 3). As the length of gestation is a relatively fixed characteristic in mammals (with their constant – homeothermic – internal environment), the breeding time controls the time of parturition. In the cold-blooded (poikilothermic) viviparous vertebrates (certain snakes and fish, for instance), the length of gestation is more variable and largely depends on temperature. For most mammals it is usually assumed that the time when the animal is most vulnerable to detrimental environmental effects is at birth, and therefore that the time of breeding has been determined by natural selection so that births occur when the environment is optimal. This assumption may not be valid in some species. Recent investigations have shown that the late stage of lactation puts a considerable drain on maternal nutrition because the demands of the young, now large and growing fast, are greatest at this time. And so late lactation may well be much the most crucial period, especially for mammals living in areas of poor food supply. Ewes mate in the autumn and start lambing early in the new year, when late snows make life hard for the young lambs, but this timing means there will be ample pasture to support the needs of late lactation.

DELAYS OF REPRODUCTION

Two major physiological modifications of gestation have arisen in certain mammals to lengthen the period from conception to parturition. These are delayed implantation and delayed fertilization. In *delayed implantation*, fertilization and development of the early embryo follow copulation, but progress stops at the stage of the blastocyst, and the embryo fails to implant in the uterus; it may remain thus, in suspended animation as it were, for an extended period, resulting in a lengthened gestation. The behaviour of the blastocyst of the European badger is a little different in that it does not quite come to a stop but continues to increase in size very slowly during the delay period. Implantation of the suspended embryo may be tied to a particular time of the year (seasonal) or may occur any time (aseasonal). Aseasonal delayed implantation is found in laboratory rodents where heavy lactation is responsible for the delay. This state of affairs has been recorded for only one wild rodent, the tree mouse (*Phenacomys longicaudus*). It is also found in many continuously breeding marsupials (the macropods – kangaroos and wallabies) where a blastocyst results from mating at post-partum oestrus but does not implant while a joey (pouched young) is suckling. As implantation in marsupials is strictly speaking rather different from other mammals, delayed implantation is now called *embryonic diapause* in this group.

Examples of seasonal delayed implantation are shown in Fig. 4-6. With the exception of a single species of bat and another of deer, it is found only in some edentates (sloths, armadillos, anteaters), two carnivore families, and in seals. The phenomenon appears to be very widespread in these four groups. In mustelids (polecat, mink, stoat, badger) a spring or summer mating is followed by a prolonged delay of implantation, but birth occurs fairly uniformly in the spring. Midsummer fertilization in bears results in birth during midwinter while the females are in a state of dormancy. In seals both fertilization and birth are concentrated in midsummer when the species are hauled out on the mating beaches. In all cases of seasonal delayed implantation, the young are born at the optimal period of the year. In the larger species implantation occurs in winter, an inopportune time for the males to be fertile, suggesting that the delay period has evolved to allow fertilization earlier in the year.

Delayed fertilization is found only in bats – it has been identified in two genera, *Eptesicus* and *Myotis*. The animals copulate in September and October, although spermatogenesis has been maximal a month earlier. Follicular growth has occurred in the ovary but there is no ovulation at this time. Instead, the spermatozoa are stored immotile either in uterine crypts or in the upper vagina, and then both sexes enter hibernation. When the female emerges from hibernation, eggs are ovulated, spermatozoa become motile and fertilization takes place. Young are born in early summer. A peculiar feature of this type of reproduction is that in some species males also copulate with females on emergence from hibernation, but the males are apparently then infertile.

Delayed development (extended gestation) is found also in two species of bats. In the California leaf-nosed bat (*Macrotus californicus*) the blastocyst implants normally but the embryo grows very slowly for a 4-month period. *Miniopterus schreibersii*

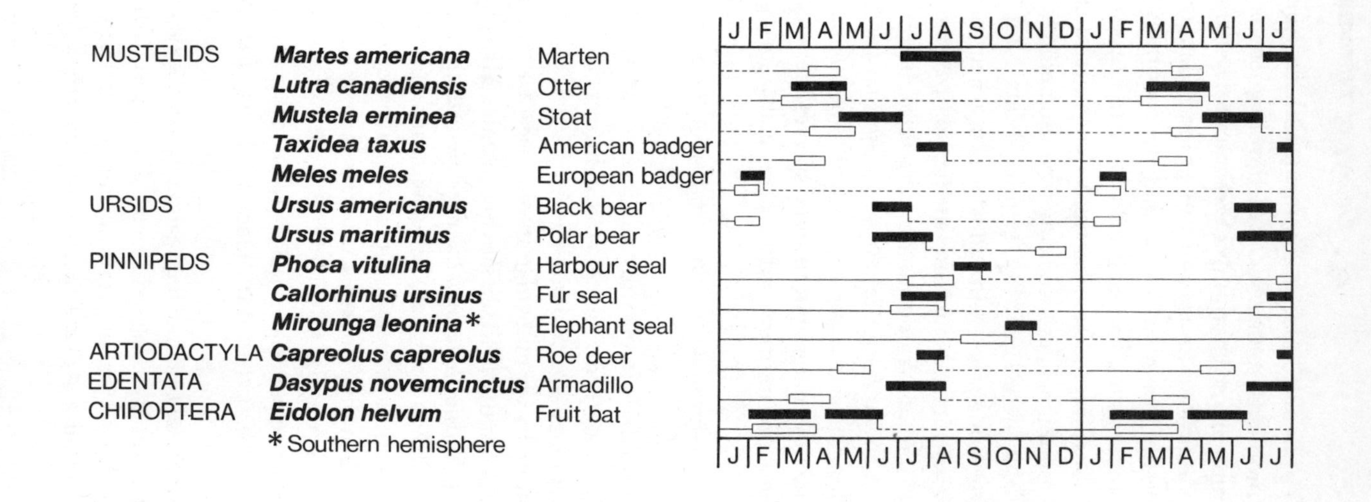

Fig. 4-6. Breeding patterns of some mammals with delayed implantation.

mates and the egg is fertilized in autumn (southern Europe). The blastocyst implants soon after, but development is very slow and the young are not born until early summer at the end of hibernation. Gestation is thus from 3 to 4 months longer than in tropical species of the same genus, and the low body temperature of the hibernating animal seems to be the reason.

In many hibernating rodents, particularly squirrels in the genus *Citellus*, spermatogenesis starts before the animals enter hibernation or just afterwards, but is then delayed for some months until just before the end of hibernation when the male gonads and accessory glands increase further in size. Similarly ovarian follicles start enlarging in the middle of hibernation. The animals copulate as soon as they emerge so that the breeding season is directly related to the end of hibernation. The bears are not true hibernators but in some species the young are born in the overwintering den and emerge in the spring with their parents.

The interrelationships of breeding and migrations are well known in birds and some fish; in mammals migration is less important, except in marine forms such as seals and whales. Caribou (North American reindeer) migrate in large herds northwards from the forest to the tundra. The females give birth in June immediately after migration ceases, having travelled from 300 to 400 miles while in late pregnancy! Many whales give birth in temperate or tropical waters just after they have migrated from polar regions, and seals haul out on traditional breeding beaches after migrating to them from considerable distances. They then give birth almost immediately.

This chapter has been concerned with diverse patterns of breeding in mammals. They reproduce successfully from the Arctic tundra to equatorial jungles, from the Australian desert to the wet west coast of Canada. Their production ranges from a single young per female every three or four years, to twenty or more every year. These features demonstrate the high adaptability of mammalian reproductive processes. Special mechanisms (post-partum oestrus, delayed implantation, delayed fertilization) allow mammals to reproduce efficiently in environments with detrimental seasonal characteristics, but we know little of the factors that govern selection of reproductive mechanisms. Energy costs of reproduction must play a major role in determining whether or not a species is monoestrous or polyoestrus, whether copulation occurs in spring or autumn, or whether its breeding is restricted or continual. It is in this new field of reproductive energetics that many of the answers to the question of 'why does such-and-such a species reproduce in such-and-such a way' will be found.

SUGGESTED FURTHER READING

Patterns of Mammalian Reproduction. S. A. Asdell. London; Constable (1965).

Vertebrate Sexual Cycles. W. S. Bullough. London; Methuen (1951).

Reproductive Behaviour in Ungulates. A. F. Fraser. London; Academic Press (1968).

Comparative Biology of Reproduction in Mammals. Ed. I. W. Rowlands. Symposium No. 15 of the Zoological Society of London (1966).

Biology of Reproduction in Mammals. Ed. I. W. Rowlands and J. S. Perry. Supplement No. 6. *Journal of Reproduction and Fertility* (1969).

The Ecology of Reproduction in Wild and Domestic Mammals. R. M. F. S. Sadleir. London; Methuen (1969).

Reproductive Physiology of Vertebrates. A. van Tienhoven. Philadelphia; Saunders (1968).

5 Fertilization

C. R. Austin

New life begins at fertilization, with the union of two germ cells, the egg and the spermatozoon. In mammals the germ cells arise in different individuals, as described earlier in this book. In the present chapter we shall take up the story from the moment of release of the germ cells from the organs of their production, the ovary and testis, and consider some of the complex machinery responsible for bringing them together, and for promoting the act of fertilization. In doing so we shall be concerned mainly with anatomy and physiology, and with the cellular features and hereditary significance of fertilization; in other chapters, there are discussions on more general aspects, controlling mechanisms and indirect influences. The consequences of errors in fertilization are discussed more fully in Book 2, Chapter 4.

The focal point for the events we are now concerned with is the 'site of fertilization' – the primary destination of eggs and spermatozoa, and the locale for the initiation of development. In nearly all mammals the site is in the ampulla, a wide part of the oviduct or Fallopian tube. Exceptions are found in some rare insectivores in which fertilization regularly occurs in the ovarian follicle soon after rupture. This may occasionally happen also in other mammals; in man, it could account for ovarian pregnancy, when the embryo grows for a time attached to this organ. After fertilization in the ampulla, the egg – or more correctly the zygote, and later the embryo – passes through the rest of the oviduct into the uterus where by far the greater part of development takes place.

EGG TRANSPORT

In earlier chapters we saw how the egg develops in an ovarian follicle, and how its nucleus goes through several changes, usually just before ovulation, which make up the first meiotic division and the beginning of the second. In most mammals, then, the egg is ovulated in the metaphase stage of the second meiotic division: in the dog and fox the egg is released just before meiosis begins. When meiosis is ultimately completed, the egg will have half the number (the haploid complement) of chromosomes, and also the reduced amount of the DNA, as compared with the content of most normal tissue cells in the animal, which are diploid; this reduction is brought about by the discarding of chromosomes in both the first and second polar bodies.

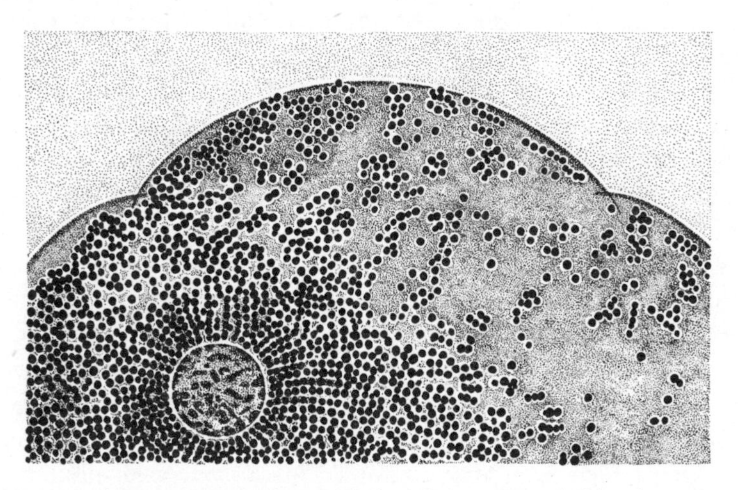

Fig. 5-1. A rat egg soon after ovulation; it is surrounded by the cumulus oophorus, consisting of follicle cells embedded in a clear gelatinous matrix. (C. R. Austin and E. C. Amoroso, *Endeavour*, **18,** 130, Fig. 35 (1959).)

The ovulated egg retains its covering layers of follicle cells and the gelatinous material in which they are embedded – collectively termed the cumulus oophorus – and immediately enters a new environment (Figs. 5-1 and 5-2). In rodents and some other mammals such as the ferret, dog and cat, the ovary is surrounded by a complete, or almost complete, tissue capsule

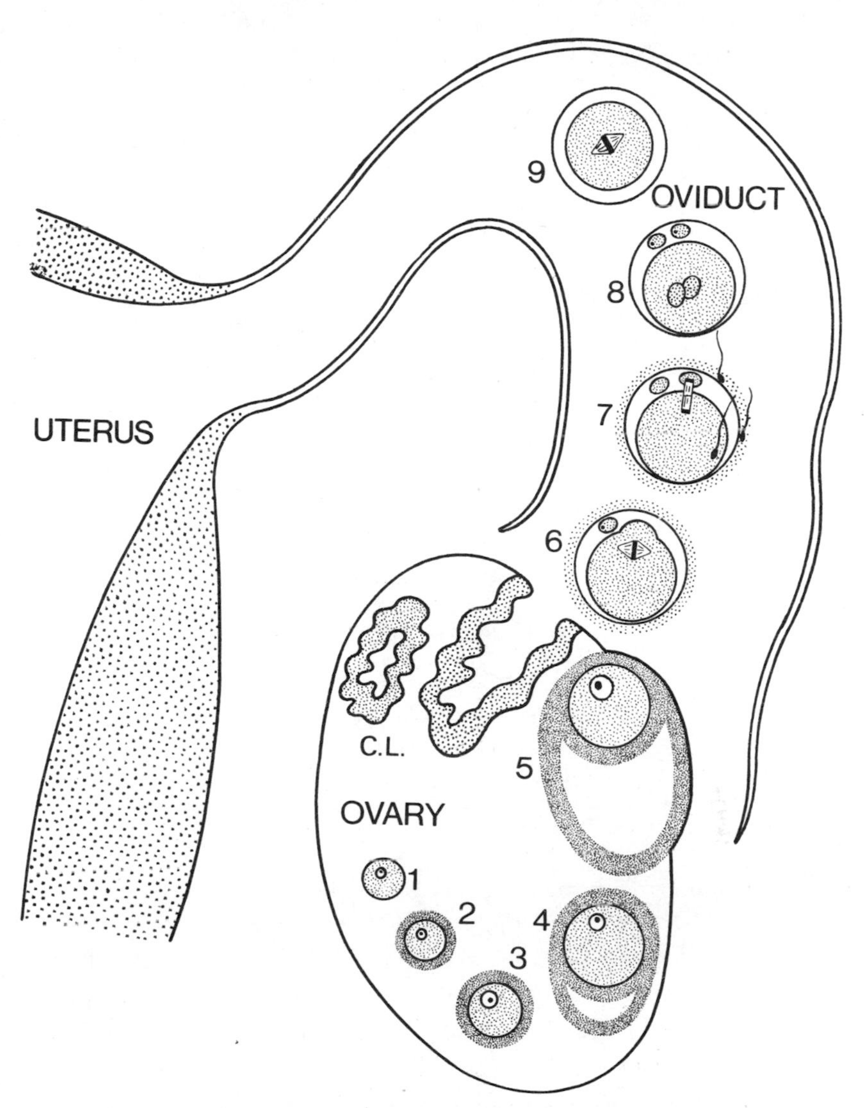

Fig. 5-2. Diagram of the human ovary, oviduct and part of the uterus, showing stages in the development of the egg and early embryo. 1, primordial oocyte; 2–4, formation and growth of follicle; 5, Graafian follicle; 6, recently ovulated oocyte with first polar body and second maturation spindle; 7, sperm penetration; 8, pronuclear development; 9, fertilized egg with first cleavage spindle. CL, corpora lutea. (Modified from Tuchmann-Duplessis, G. David and P. Haegel, *Illustrated Human Embryology*, vol. 1. New York; Springer-Verlag, London; Chapman and Hall, Paris; Masson et Cie (1971).)

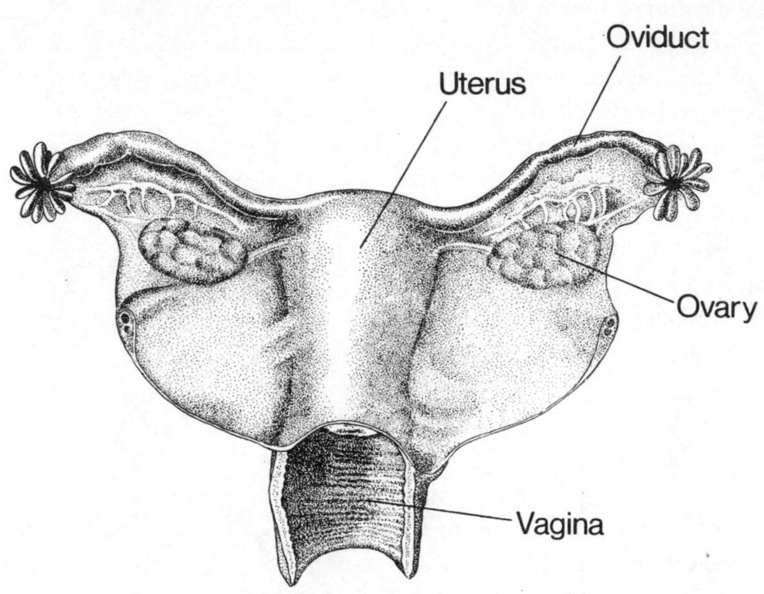

Fig. 5-3. The human uterus and associated structures, the oviduct, the ovary, and part of the vagina opened to show the interior. The finger-like projections of the fimbria are clearly seen at the free ends of the oviducts. (J. T. Velardo Ed. *Endocrinology of Reproduction*. Oxford University Press (1958).)

continuous with the oviduct, and so the eggs can hardly fail to find their way from the capsule into the oviduct. They are moved in the required direction by currents in the fluid secretions filling capsule and oviduct, the currents being produced by the vigorous beating of vast numbers of cilia on the oviduct walls. In other mammals, the tissues about the ovary form, to varying degrees, a kind of capsule, but in many species nothing worthy of the name exists and eggs seem liable to miss the oviduct and fall into the body cavity. Such is the state of affairs in rabbit and man (Fig. 5-3), but in fact the eggs rarely go astray and their normally successful transfer from follicle to oviduct is owing to the diligent manoeuvrings of the infundibulum. This is the name of the funnel-shaped free end of the oviduct, the edge of which is developed as a fringe, the fimbria, with finger-like projections richly endowed with cilia. As a result of the movements of the infundibulum over the surface of the ovary at the time of ovulation, the eggs are captured and carried by the cilial current into the depths of the oviduct. Once in the oviduct transport is both by cilial currents and muscular contractions of the oviduct walls. Both the rate of cilial beat and the vigour of the muscular contractions are controlled by the prevailing balance of the sex hormones, oestrogen and progesterone.

Thus the eggs reach the site of fertilization, and they generally do so to find the spermatozoa waiting for them, because in most mammals the period of sexual receptivity, oestrus, begins several hours before ovulation. Uniquely in the human subject there is no fixed relation between coitus and ovulation, and sometimes eggs must await the arrival of spermatozoa. The egg's fertile life is comparatively brief – probably under 24 hours except in a few species such as the horse, in which it may last for 2 or 3 days. After the first few hours, the waiting egg begins to deteriorate or age, a change that can seriously affect later embryonic development. The consequences of egg ageing are discussed in more detail further on in this chapter and also in Books 2 and 4.

The egg-transport system can break down at several points. As we have seen, the egg may fail to leave the follicle, and undergo fertilization and some development there, producing ovarian pregnancy, one form of extra-uterine or ectopic pregnancy. Or it may pass into the peritoneal cavity (possibly after fertilization in, or shortly after leaving, the follicle) and attempt to develop by attaching to the surface of one of the organs there (abdominal pregnancy). The commonest form of ectopic pregnancy occurs when the egg enters the oviduct and undergoes fertilization, but is not moved on and instead becomes attached to the wall of the oviduct, setting up a tubal pregnancy. Ectopic pregnancies are most common in man, and usually do not go far, though abdominal pregnancy can go to term (requiring then Caesarian delivery).

SPERM TRANSPORT

The transport of spermatozoa from the testis to the site of fertilization is indeed quite a saga. Leaving the testis they pass into an exceedingly long tube twisted into a compact body called the epididymis, lying next to the testis (Fig. 5-4). The

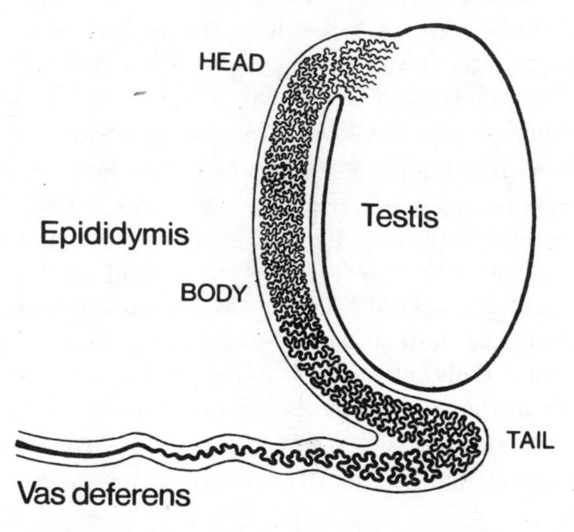

Fig. 5-4. Diagram of the testis and epididymis, to show how the fine tubules that leave the testis join up to form a single highly convoluted tube which eventually becomes the vas deferens.

contents of the epididymis consist of enormous numbers of spermatozoa suspended in secretions of testis and epididymis, and this mixture moves slowly down the tube. Anatomically we distinguish three regions of the epididymis, labelled simply head, body and tail; the spermatozoa pass through each in turn, finally entering the continuation of the epididymal tube, a conduit of somewhat wider dimensions called the vas deferens. Some of the suspending fluid is absorbed in the epididymis so that the concentration of spermatozoa in the vas deferens is remarkably dense, the cells literally rubbing shoulders with one another. Traversing this structure the spermatozoa enter the urethra, the duct that carries the urine from the bladder.

When the spermatozoa leave the testis they are neither structurally nor functionally complete, and have still to undergo several changes as they move through the epididymis. There is a change in the shape of the head, and in particular in the form of an important cap-like component, the acrosome, which becomes somewhat reduced in size. On leaving the testis, the spermatozoon carries a small cytoplasmic droplet attached to the midpiece, and this is normally discarded during passage through the epididymis. If an animal ejaculates very frequently, an increasing proportion of the spermatozoa in the semen are found with cytoplasmic droplets, and this is taken as a sign that the sperm supply is being called upon too quickly for the normal maturing process to take place in the epididymis. Functional change is shown experimentally by the fact that spermatozoa taken from the head of the epididymis are infertile when introduced into a female of the species by artificial insemination. Only after they have passed through most of the body of the epididymis are the spermatozoa fertile with artificial insemination, but whether this functional change is linked in some way to the structural changes mentioned is unknown.

Spermatozoa leave the male tract at ejaculation, provoked by masturbation or copulation or in spontaneous emissions, or failing these events they pass out in a slow steady discharge that is carried away by the urine. In ejaculation, the spermatozoa from the vas deferens and from the tail of the epididymis are rapidly forced into the urethra by muscular contractions of the duct walls, mingling with the secretions of certain accessory glands whose ducts lead into the terminal part of the vas deferens or into the urethra. The male accessory glands are developed to different degrees in different species. In man and several other mammals these are the ampullary glands, the seminal vesicles, the prostate, the urethral or Littré's glands, and the bulbourethral or Cowper's glands (Fig. 5-5). The pig has no ampullae, the dog and cat lack seminal vesicles, the bull has large seminal vesicles but small prostate, and monotremes and some marsupials have little else but bulbo-urethral glands. In rodents there are so-called coagulating glands which can be considered as anterior developments of the prostate gland. The contributions from these sources with the secretions from testis, epididymis and vas deferens, and the suspended spermatozoa, together constitute the semen. The seminal fluids (plasma) serve primarily as a vehicle, but in addition some components stimulate the activity and metabolism of the spermatozoa, and others provide the necessary energy. Characteristic chemical constituents of seminal plasma normally present only in traces in other bodily secretions, include such things as fructose,

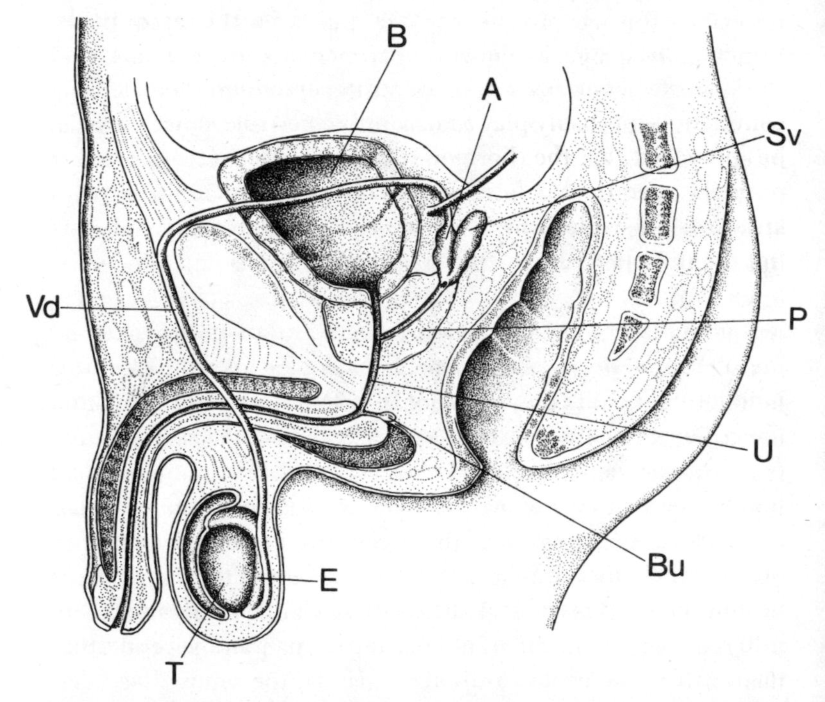

Fig. 5-5. Diagram of the human male reproductive tract, and neighbouring organs. A, ampulla; B, bladder; Bu, bulbo-urethral or Cowper's glands; E, epididymis; P, prostate; Sv, seminal vesicle; T, testis; U, urethra; Vd, vas deferens. (Littré's glands are very small and not shown here.)

sorbitol, inositol, citric acid, phosphorylcholine, ergothioneine, and glycerylphosphorylcholine.

At copulation (coitus) the semen in some species is deposited directly into the vagina, as in man, sheep, cow and rabbit, or may be forced through the cervix into the uterus (mare, sow and rodents). To different degrees in different species the semen coagulates after ejaculation and this is thought to be important in retaining the spermatozoa and preventing their loss from the female tract. In rodents like rats and mice the coagulum is especially large and tough, forming a 'copulation plug' which fills the vagina and may persist for one or two days. The mechanical factors responsible for propelling the semen into the female tract include muscular contractions in the wall of the urethra, the piston-like action of the penis during coitus, and the orgasmal contractions of the female tract. The female tract contractions probably play a fairly prolonged role, since they can be provoked by the hormone oxytocin from the posterior pituitary gland; also, semen contains a group of substances known as prostaglandins which are capable of strongly stimulating the muscle tissue in the walls of the tract.

By the means described, spermatozoa accompanied by differing amounts of seminal plasma achieve the further reaches of the uterus, near its junction with the oviduct (the utero-tubal junction). The junction in many species has a narrow bore and the region evidently represents a partial barrier to sperm transport. Spermatozoa that do manage to negotiate the utero-tubal junction are carried along the oviduct chiefly by the irregular contractions of the walls of the organ, which cause a mixing of the contents; in addition there are on the walls bands of cilia beating in the direction of the ovaries, and consequently there is to some extent a current of fluid favouring passage of spermatozoa to the site of fertilization.

To reach their destination, spermatozoa do not appear to depend upon their own swimming movements. Observed under the microscope, spermatozoa generally swim in wide circles, and there is no sound evidence of anything in the female tract that would give them a sense of direction. Moreover, the time taken to reach the site of fertilization – often only five minutes after coitus – is much too short for them to swim the distances involved. The last point may not apply to the rabbit with its unusually long sperm transport time – 3 to 4 hours – but, for most animals, one may well question the value of the swimming movement of the spermatozoon. Probably it is important in two ways: firstly it would help to keep the cell in suspension and thus facilitate its transport by the mechanisms that we have considered, and secondly it is probably necessary for the transit of individual spermatozoa through the coats that enclose the egg, the cumulus oophorus and zona pellucida – of these things more later.

Passage to the site of fertilization is not an easy task for spermatozoa; we have seen that the utero-tubal junction provides a partial barrier, and the same is also true for the uterine cervix when ejaculation is directly into the vagina. In addition, spermatozoa are subject to considerable dilution in the secretions of the female tract, and the machinery responsible for their transport seems curiously crude. The system certainly has a low order of efficiency, in terms of the proportion of spermatozoa that reach their goal. While tens of millions or even hundreds of millions of spermatozoa are deposited in the vagina (or, in the mare, sow and rodents, in the uterus), the numbers reaching the site of fertilization rarely exceed a hundred in the small animals, and a few hundred to a few thousand in the large. The magnitude of sperm numbers at the site of fertilization may still appear well in excess of needs, but in fact the population is really very sparse when one takes into account the minute size of the spermatozoon and the comparatively enormous volume of space within this region of the oviduct.

Contemplating the relative scarcity of spermatozoa at the site of fertilization on the one hand and the high efficiency of fertilization on the other, people have often been tempted to invoke a chemical attraction of spermatozoa by the eggs, a mechanism referred to as chemotaxis, which is known to aid the meeting between egg and spermatozoon in the simpler members of the plant kingdom such as the algae, mosses and ferns. There is also evidence that a like attraction plays a role in bringing about the assembly of spermatozoa and eggs in some primitive marine animals. But, despite much experimentation and careful observation, no good reason exists for thinking that mammalian germ cells are chemically attracted to each other, and we must infer that their eventual contact at the site of fertilization is controlled by chance.

Like the eggs, spermatozoa have quite a short fertile life span in the female tract – generally, not more than 24 hours though there are some striking exceptions. In the domestic hen, spermatozoa find their way into somewhat specialized crypts in the wall of the oviduct. Numbers emerge as each egg passes, and become involved in fertilization, but the supply is conserved in some strange way and often lasts for 3 weeks or more. In certain bats coitus takes place in the autumn, and the animals then retire for their winter sleep which lasts 3 or 4 months; when they wake up, ovulation takes place and fertilization by the autumn spermatozoa ensues. Otherwise spermatozoa must complete their task within a few hours if the outcome is to be normal, for they deteriorate with the passage of time, and this reduces the possibilities of healthy embryonic development.

Nature's use of spermatozoa seems a prodigal affair in mammals. In the honey bee, by contrast, the female tract methodically rations out one or a few spermatozoa for each egg that is to be fertilized, and thus the contribution received on one nuptial flight lasts several years (a tribute also to the longevity of the spermatozoa). But in mammals, of the tens or hundreds of millions of spermatozoa deposited in the female tract, only one or a few, depending on the number of eggs fertilized, are actually involved in fertilization. This poses the questions: what happens to the others, the vast majority? Have they any significance? We can answer the first with confidence, but the second is open to debate. Most of the spermatozoa left in the uterus are passed into the vagina, and voided from there in vaginal secretions or with the urine flow. Those in the oviducts and some of those remaining in the uterus are consumed by the phagocytic white cells that invade the tract in the second half of the oestrous or menstrual cycle. In the oviducts, some spermatozoa are phagocytosed by epithelial cells lining these organs. Despite these disposal mechanisms a few spermatozoa may still be around (though doubtless no longer alive) when the embryo arrives on the scene, and these may be phagocytosed by the embryonic trophoblast cells.

As to the significance of the residual spermatozoa, the suggestion has often been made that, as little masses of foreign protein, they might cause some immunological response in the female, or that, as quanta of genetic information, they might influence the future functioning of the female tract tissues or even the embryo. The former supposition is evidently true, since anti-sperm antibodies have been demonstrated in both serum and tract secretions in animals that have mated, but this reaction rarely influences female fertility (see Book 4, Chapter 4). There is really no direct evidence at all on possible genetic effects.

CAPACITATION, ACROSOME REACTION AND SPERM PENETRATION

Although spermatozoa that have passed through the epididymis are fertile if introduced into the female tract by coitus or artificial insemination, they must undergo still further changes before they can actually engage in fertilization. These changes normally take place in the female tract, though as we shall see later conditions can often be set up for their occurrence *in vitro*. The first change is known as capacitation, and this is followed by a structural alteration in the acrosome identified as the acrosome reaction. The nature of capacitation is unknown, but readily repeatable (and often repeated) experiments have shown that ejaculated spermatozoa must spend a definite period of time in the female tract, or under specified conditions *in vitro*, before becoming able to penetrate the coats enclosing the egg (Figs. 5-1 and 5-6). The time required for capacitation varies in different animals: in the rabbit it takes about 5 hours after coitus, in the rat and hamster about 3 hours, and in the sheep about $1\frac{1}{2}$ hours. The rather scant evidence that we have suggests that the time for the human spermatozoon is of the order of 7 hours.

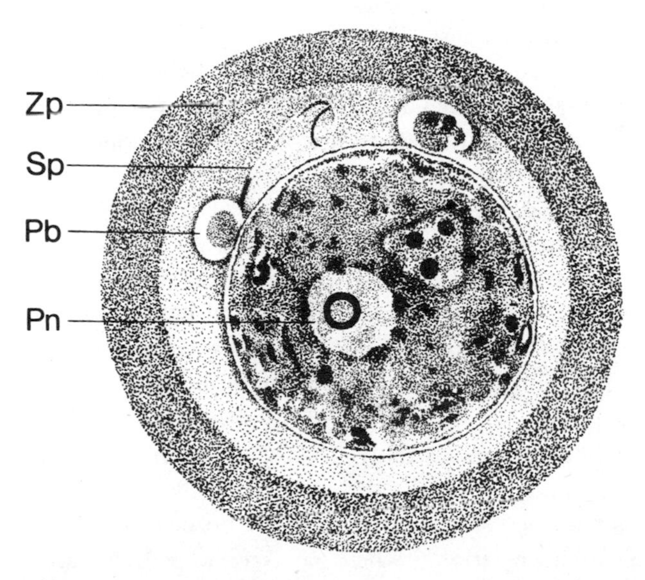

Fig. 5-6. The egg of the field vole during the pronuclear phase of fertilization. Pb, one of the two polar bodies; Pn, the male pronucleus (the female pronucleus is not quite in focus, a little above and to the right); Sp, the tail of the fertilizing spermatozoon; Zp, the zona pellucida. (C. R. Austin, *J. Anat.* **91**, 1, Plate 1, Fig. 8 (1957).)

Capacitation according to some authorities could involve an unmasking of receptor sites or perhaps the removal from the sperm surface of an inhibitory or stabilizing coat, but a surface change of this kind has not been demonstrated by electron microscopy. On the other hand, an inhibitor of some sort does seem to be deposited on the spermatozoon under certain circumstances. If we recover spermatozoa from the rabbit female tract several hours after mating, we can demonstrate that they have undergone capacitation by showing that they are then immediately capable of penetrating eggs. M. C. Chang of Shrewsbury, Massachusetts, found that when such spermatozoa were re-suspended in seminal plasma they appeared to lose this faculty, but could regain it on further residence in the female tract. The effect was attributed to a component of the seminal plasma in the nature of a mucopolysaccharide and referred to as the decapacitation factor; it has been prepared in fairly pure form and investigations on it are continuing in the hope that it might possibly provide a convenient means for human fertility control, but there are many difficulties in the exploitation of this idea.

The reason why capacitation makes it possible for the spermatozoon to penetrate egg coats appears to be quite clear – it prepares the spermatozoon to respond to substances accompanying the egg, possibly deriving from the ovarian follicle; response takes the form of the acrosome reaction. According to Ryuzu Yanagimachi in Honolulu there are actually two substances involved: one a chemical of relatively small molecular weight which preserves the vitality of the spermatozoon during its preparation, and the other protein-like which actually provokes the acrosome reaction. Spermatozoa can only undergo the acrosome reaction if they have completed the process of capacitation, and they can only penetrate the cumulus oophorus and the zona pellucida if they have undergone, or are undergoing, the acrosome reaction. The course of the reaction has now been worked out in great detail. First of all, the plasma membrane of the spermatozoon in the region of the head, where it overlies the acrosome, becomes fused at a number of discrete points with the outer membrane of the acrosome. Fusion leads rapidly to the formation of small apertures leading into the cavity of the acrosome, and so portals of exit are provided for the acrosome contents, which consist largely of hyaluronidase and a trypsin-like enzyme. The enzymes can dissolve the jelly part of the cumulus oophorus and they thus enable the spermatozoon to pass through the cumulus oophorus and reach the surface of the zona pellucida. Then the complex of fused plasma–acrosome membranes becomes detached altogether from the sperm head. The spermatozoon is now enclosed by a

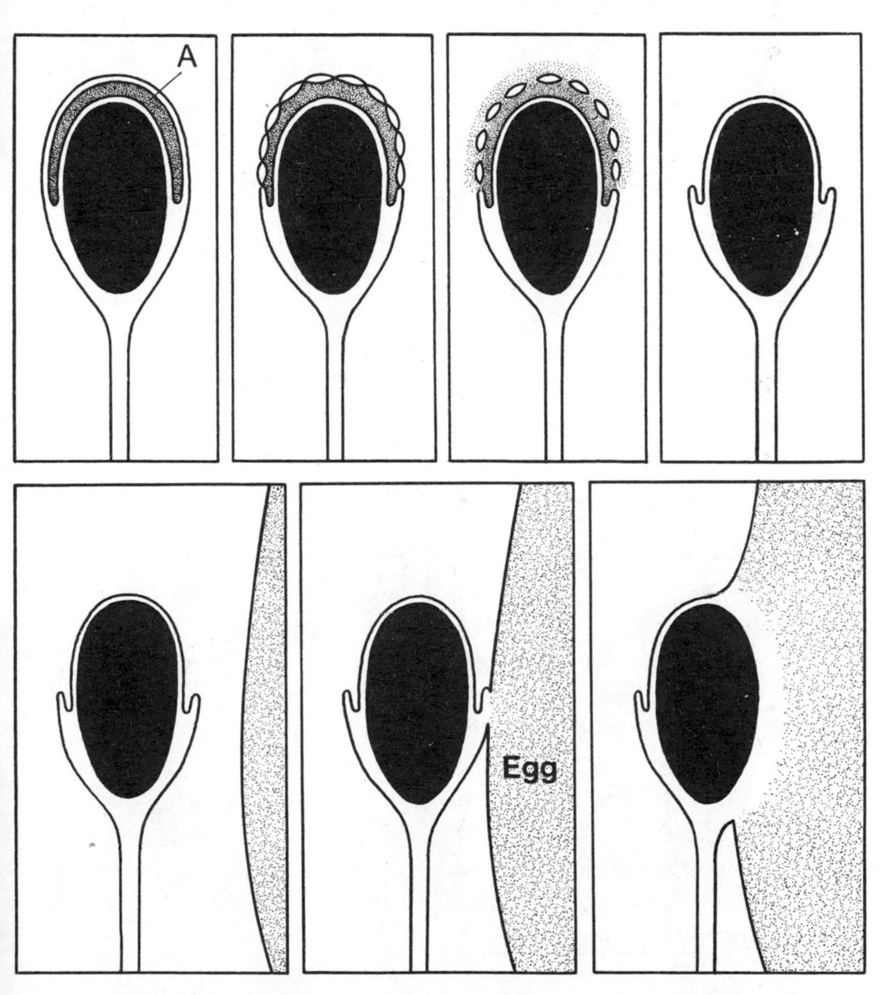

Fig. 5-7. Diagram showing the pattern of the acrosome reaction (*above*), and the first steps in sperm-egg fusion (*below*). The outline of the spermatozoon represents its plasma membrane; the nucleus is solid black. A, acrosome.

mosaic consisting of the inner acrosome membrane and the rest of the sperm plasma membrane. Figs. 5-7 and 5-8 will make it easier to follow the membrane changes.

Passage of the spermatozoon through the zona pellucida does not seem to involve actual release of enzyme, for the narrow slit that the spermatozoon makes in the zona pellucida testifies to

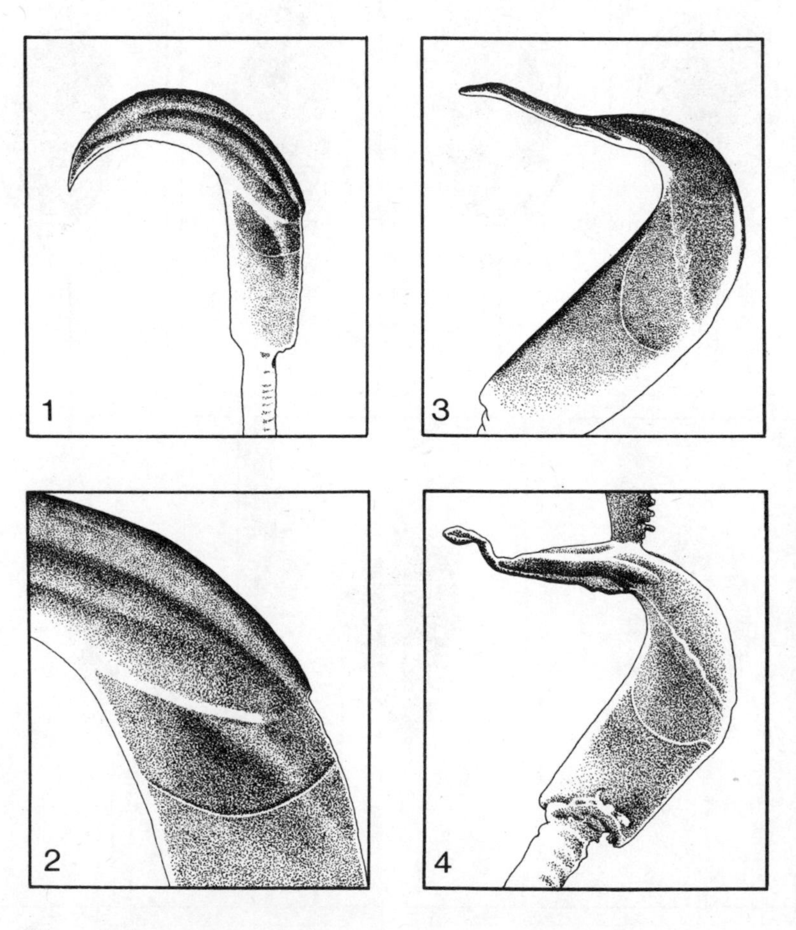

Fig. 5-8. Drawings made from photographs taken by the stereoscan electron microscope of hamster sperm heads. In 1 and 2, the sperm head is shown before reaction, and in 3 and 4 the reacted acrosome is seen partly detached. (Photographs taken by R. Yanagimachi.)

non-diffusion of lytic agent from the sperm head. We infer that the agent (termed the zona lysin) has the capacity to liquefy the zona material, and is firmly attached to the inner acrosome membrane, still disposed over the leading part of the sperm head. The spermatozoon thus enters the narrow fluid-filled space that immediately surrounds the cytoplasmic body of the egg, the perivitelline space, and is in a position to proceed with the final act of germ cell fusion.

CYTOPLASMIC FUSION

Arriving in the perivitelline space the spermatozoon attaches almost immediately to the cytoplasmic body of the egg (Figs. 5-7 and 5-9). Usually this occurs while the sperm tail is still protruding through the slit in the zona pellucida. Attachment is initially superficial but is soon followed by fusion of the

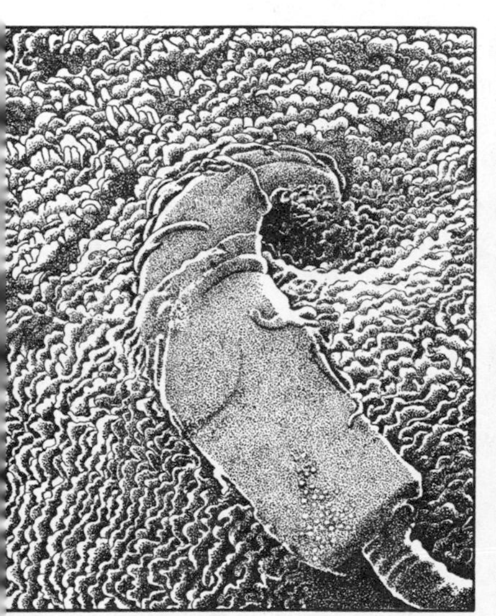
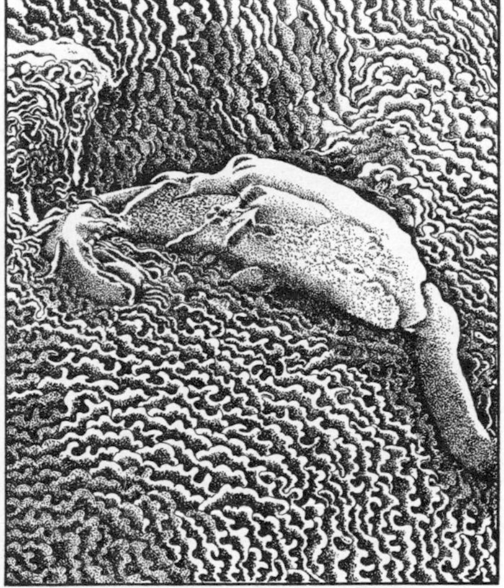

Fig. 5-9. Drawings made from stereoscan electron micrographs of hamster spermatozoa attached to, and probably undergoing fusion with, the egg surface. (Photographs taken by R. Yanagimachi.)

contiguous plasma membranes of spermatozoon and egg. This begins at the points of contact between certain small projections on the egg surface, the microvilli, and the posterior region of the sperm head. As with the acrosome reaction, membrane fusion is rapidly succeeded by perforation so that the plasma membranes of the two cells become continuous with each other, and where one plasma membrane ends and the other begins is almost impossible to say. These stages of gamete fusion were first elucidated in rat eggs and spermatozoa by Lajos Piko and Albert Tyler in Pasadena. Incorporation of the two cells within the same overall expanse of plasma membrane means that spermatozoon and egg now constitute a single cell. The bridge then progressively enlarges until the entire sperm nucleus becomes surrounded by egg cytoplasm, followed in due course by the sperm tail.

In some animals the tail of the spermatozoon is not drawn into the egg cytoplasm but left either in the perivitelline space (as commonly happens in the field vole – Fig. 5-6) or completely outside the egg (as in the Chinese hamster). The fact that passage of the tail into the cytoplasmic body of the egg is not invariable among mammals strongly suggests that the tail plays no essential role in fertilization or development. Indeed observations by electron microscopy of subsequent stages in the life of the engulfed tail support this view – the sperm mitochondria degenerate at an early stage and the other components, the coarse fibres and the axial filament of the tail, apparently undergo dissolution. Alone among the cytoplasmic parts of the spermatozoon, the centriolar complex may possibly play a further role at least in some species, in inducing a sperm aster which seems to guide the sperm nucleus in its approach to the egg nucleus – of which more later.

IMMEDIATE RESPONSE OF THE EGG TO SPERM PENETRATION

The egg reacts very rapidly to sperm entry in several ways. The basic response is that of activation, a term signifying the initiation of embryonic development. In some non-mammals the egg's response has often been shown to be quite mechanical: frog eggs pricked with a needle commonly enter upon development and often go as far as free-swimming tadpoles, occasionally to mature frogs. So the influence of the penetrating spermatozoon in activation could be a simple physical one. The current view is that the egg passes into an inhibited state when the second meiotic division reaches metaphase; its metabolic processes do continue but at a subdued level, and gradually decline as the egg ages after ovulation. Sperm entry in some way abolishes the inhibition. The underlying biochemical events of inhibition and activation have been studied in most detail in sea urchin eggs. Observations with this convenient material have shown that essentially all the components of the systems for respiration, and for protein and nucleic acid synthesis are present before activation, and that some of the earliest events involve the disinhibition of enzyme systems, and the 'unmasking' of messenger RNA and ribosomes. Before this, there is a release of proteases from little packages in the egg cytoplasm called lysosomes; these enzymes may well be responsible for the changes. Breakdown of lysosomes could be the primary effect of sperm penetration (or needle puncture). The same series of changes could take place in mammalian eggs; as yet our methods for detecting them are not sufficiently sensitive, with the relatively small number of eggs that can be got together at one time, but we do have some evidence that DNA and protein synthesis occurs during mammalian fertilization.

Other responses by the egg include both functional and structural changes: the induction of blocks to polyspermy, the opening and evacuation of cortical granules, the resumption of the second meiotic division and emission of the second polar body, and the formation of the egg nucleus. The blocks to polyspermy involve loss of permeability in the zona pellucida and alterations in the egg surface that preclude further sperm attachment, but the precise nature of the blocks remains unknown. The cortical granules (small rounded bodies lying just beneath the egg surface) open and empty about the same time as alteration takes place in the properties of the zona pellucida – perhaps the material from the cortical granules, passing across the perivitelline space, produces the change, which we refer to as the zona reaction. Perhaps too, the cortical granule response leads to the altered receptivity of the egg surface to spermatozoa. Whatever the mechanism may be, most mammalian eggs evidently have two efficient protective devices against fertilization by more than one spermatozoon. Of course these are not perfect mechanisms, and so two spermatozoa (very rarely three or more) occasionally do become fused with the vitellus and take part in fertilization. This initiates the state of polyspermy of which we shall have more to say shortly.

Resumption of the second meiotic division leads to completion of meiosis, and the egg is left with a haploid set of chromosomes; at fertilization the diploid number is restored by the introduction of another haploid set by the spermatozoon. Sometimes the second meiotic division fails to proceed in a normal way and a second polar body is not extruded; when this happens the egg comes to contain two female pronuclei and a diploid number of chromosomes. Fertilization then results in the abnormal state of polygyny, and of this, too, more later.

PRONUCLEAR DEVELOPMENT AND SYNGAMY

Soon after passage of the sperm nucleus into the egg cytoplasm, both it and the group of chromosomes remaining in the egg after the second meiotic division form distinctive nuclei, the male and female pronuclei. The initial changes in the sperm head seem to be mainly an unravelling of chromatin threads which represent the form in which the chromosomes are packed in the extremely dense sperm nucleus. This disposition, together with the association of the DNA strands with histone, a gene-repressing protein, probably ensures metabolic inertness in the nucleus of the intact spermatozoon. The more picturesque phase of fertilization now ensues, with the growth and enlargement of the two pronuclei (Figs. 5-10 and 5-11). These are striking structures because of the nucleoli they contain – highly refractile spheres of different sizes, varying in number from one to thirty or more. The nucleoli make their appearance as the pronucleus itself is formed, and increase in size and number as the pronucleus grows. The pronuclei achieve a remarkable size, each much larger relatively than the nucleus in most tissue cells, and as they reach their full dimensions they move towards each other and come into contact, fairly precisely in the centre of the egg.

In mammals, all this is quite a stately performance, taking about 12 hours, in contrast to the 1 hour that the pronuclear phase involves in the sea urchin. There is the further difference that in the sea urchin the male pronucleus actually fuses with the female pronucleus, the two masses of chromatin becoming enclosed within the same expanse of nuclear envelope and forming what is known as a zygote nucleus. In mammals, the two pronuclei, after being in contact for a while, show a further series of changes, ushering in the process of syngamy without at any time fusing together. First there is a gradual shrinkage in size and a diminution in the number of nucleoli within the two pronuclei. Then the nuclear envelopes break up and disappear, and the remaining nucleoli fade out until little remains in the positions once occupied by the two pronuclei, other than two vague gatherings of partially condensed chromosomes. The chromosome groups complete their condensation and assemble together – at last uniting the maternal and paternal hereditary contributions – in what is at once the consummation of syngamy, the closing scene of fertilization, and the prophase of the first cleavage division. This is the zygote stage in mammalian development; the next step will be the first cleavage division of the embryo.

During the course of pronuclear growth DNA is synthesized from cytoplasmic precursors, and by the time the pronuclei have reached their final size they have each doubled their

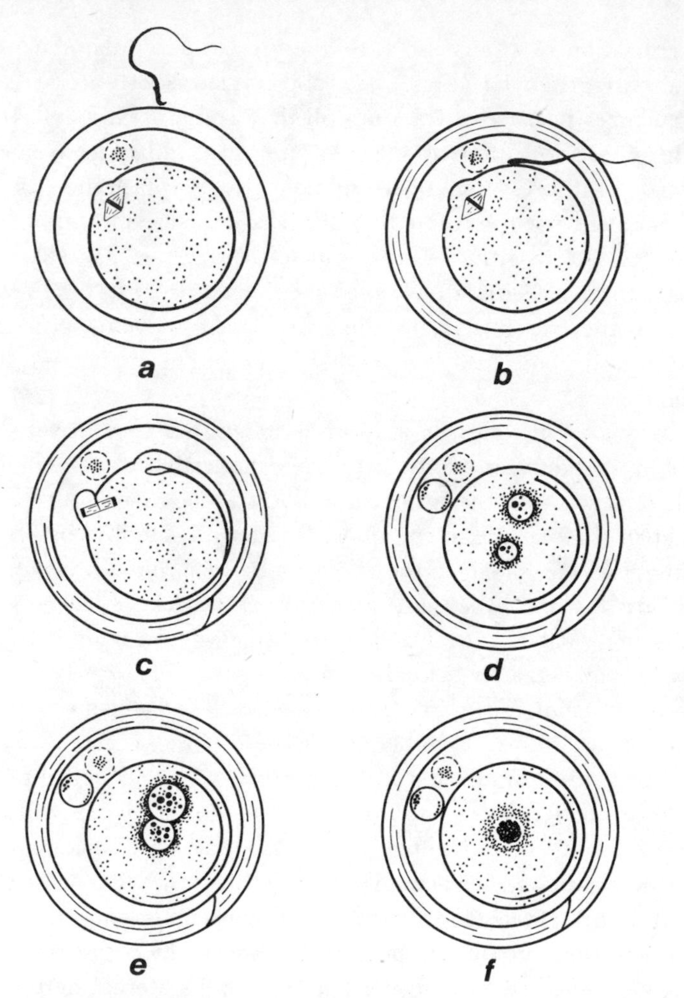

Fig. 5-10. Diagrams of the general course of fertilization, as seen in the rat egg.

a Approach of spermatozoon.

b Sperm–egg fusion.

c and *d* Emission of second polar body and formation of male and female pronuclei.

e Syngamy of fully grown pronuclei.

f Assembly of chromosome groups.

(C. R. Austin and M. W. H. Bishop, *Biol. Rev.* **32,** 296, Text-fig. 1 (1957).)

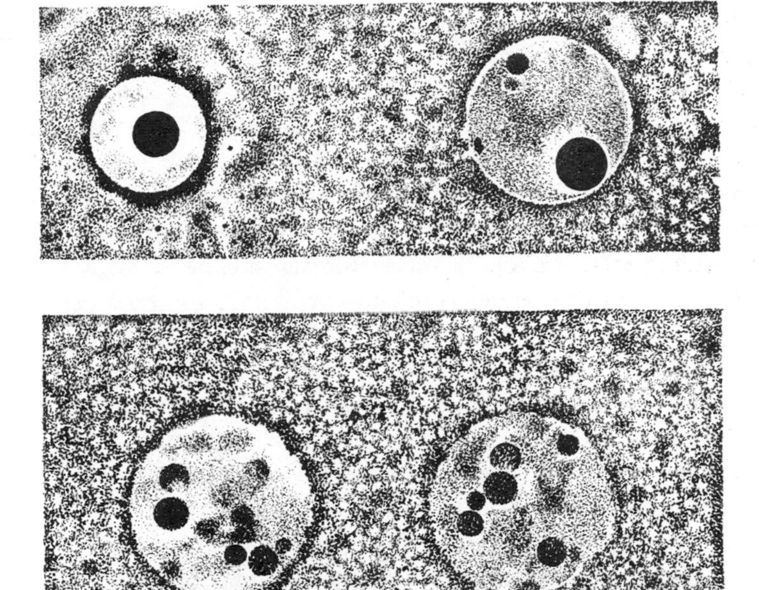

Fig. 5-11. Pronuclei in eggs of two rodents, the Libyan jird (*above*) and golden hamster (*below*). (C. R. Austin and E. C. Amoroso, *Endeavour*, **18**, 130, Figs. 31 and 32 (1959).)

original content. The nuclei formed by the first cleavage division are diploid and their DNA content is equal to that characteristic for somatic cells of the species. In contrast to the DNA synthesis, no RNA synthesis can be detected during fertilization, and the protein synthesis that we referred to earlier must depend upon 'long-lived' messenger RNA formed in the oocyte before ovulation.

Fertilization not only establishes the diploid state for the embryo but is also the time at which the sex of the embryo is decided. The egg begins by having two X-chromosomes and loses one into a polar body during meiosis; the normal complement is made up by the spermatozoon which introduces either an X- or a Y-chromosome, the result being a female or male embryo, respectively. In mammals, the female is the homogametic sex because it produces homogeneous gametes or germ cells (all having X-chromosomes) while the male is the heterogametic sex with either X- or Y-chromosomes. The situation is different in other animals; for example, in birds and some amphibians and fish the male is the homogametic sex with two like chromosomes (identified as ZZ in this case) and the female is heterogametic (ZW). In these animals, therefore, the sex of the new individual is determined at ovulation, and thus before fertilization.

To control the sex of offspring in mammals by separating the X- and Y-bearing spermatozoa has long been an ambition. There have been many claims for success, but none has been substantiated, and recent work on spermatogenesis (see the third chapter of this book) provides evidence that X and Y spermatozoa are unlikely to differ in form, and are therefore probably not separable experimentally. On the other hand, investigators have observed that the Y-chromosome shows a distinctive yellow glow (fluorescence) when tissue cells are treated with the dye, quinacrine, and it is possible that the method might be adapted to identify the Y-bearing spermatozoa.

ERRORS OF FERTILIZATION

We have already had occasion to refer to two errors of fertilization, namely polyspermy and polygyny. In *polyspermy* two spermatozoa (very rarely more) enter the egg and both take part in fertilization. The general course of events then resembles normal fertilization, except that there are now three pronuclei, two male and one female, and they do not grow as large as in normal fertilization. At syngamy all three pronuclei come together, undergo diminution, and give place to three chromosome groups which move together for the prophase and then the metaphase of the first cleavage division. The zygote thus comes to possess a threefold chromosome complement, and is described as being triploid. The first cleavage spindle looks quite normal, despite its excess of chromosomes, and the division proceeds without hitch. In fact the embryo continues to nearly half way through pregnancy just as if it were normal and diploid in chromosome content. About mid-term, however, something goes awry, precisely what we do not know, and the embryos degenerate and die. Polyspermy thus leads to triploidy and this is a lethal condition.

The behaviour of the mammalian egg in polyspermy appears to be unique among animals. In species with large eggs, such as birds and reptiles, the problem of excluding supernumerary spermatozoa over a vast expanse of surface area has evidently proved sufficiently forbidding for them to have adopted another strategem – the extra spermatozoa are allowed in and form male pronuclei, but only one of these is permitted to pair with the female pronucleus. The other male pronuclei (often fifty or more) are bundled out of the way and play no part in fertilization. This system is known as physiological polyspermy. In all other animals polyspermy is pathological, but the patterns of events vary a good deal. In many species with eggs about the same size as those of mammals, such as the sea urchin, surface barriers are erected to keep out extra spermatozoa; when protection fails, fertilization continues but the first cleavage division is chaotic owing to the formation of a mitotic spindle with three, four or more poles. These embryos are grossly abnormal from the start and never go far. In other animals, such as the frog, polyspermy results in the induction of two or more first cleavage spindles, and again development is so aberrant it is doomed to an early end. Apparently only in mammals is cleavage normal after polyspermy, and fairly extensive development possible.

The story with *polygyny* in mammals is similar, except that we now have two female pronuclei owing to failure of emission of the second polar body, and these develop along with the single male pronucleus (assuming penetration by only one spermatozoon). Subsequent events are also similar, with the formation of a triploid embryo and its death around the mid-point of pregnancy. (Further information on triploid development is given in Book 2, Chapter 5 and Book 4, Chapter 5.)

Among certain fish known as Amazon molly, there is a species that displays a curious feature in its reproductive pattern – there are no males! The females mate with males of a bisexual species of the same genus. But spermatozoa do not take part in the full course of fertilization: they penetrate the eggs, thus activating them, but fail to form pronuclei and merely degenerate. The activated eggs of this fish are capable of proceeding to full development; they refrain from emitting a second polar body and so are diploid without a contribution from the spermatozoon. This mode of reproduction is termed *gynogenesis*; it is normal in one or two other non-mammalian animal forms also. One occasionally meets examples of rudimentary gynogenesis in mammalian eggs, but there is no evidence that it can go far; this is presumably because the mammalian egg rarely manages to suppress its second polar body, and accordingly is left with only a single set of chromosomes (a state known as haploidy), which is inadequate for proper development in mammals.

The complementary process to gynogenesis, namely the failure of the female pronucleus and the initiation of development with only a male complement of chromosomes is termed *androgenesis* and is not regarded as normal in any animal form. Now and then one encounters mammalian eggs undergoing initial androgenesis, but again there is no reason to suppose that extensive development is possible.

The final variety of fertilization anomaly that we should consider here involves a rather curious chain of events called *immediate cleavage*. In some mammals, apparently including man, an early consequence of egg ageing is a drift of the second maturation spindle from its normal position just under the surface of the egg to a point more or less in the centre. The second meiotic division then proceeds spontaneously, but since the spindle is now far from the surface of the egg it cannot form a normal kind of polar body, and instead divides the entire egg into two more or less equal halves. One half is the egg and the other the polar body, but they cannot be told apart. Both halves can undergo fertilization.

These events have only been seen in the eggs of some laboratory animals, and even here we do not have direct information on the outcome, but some speculation is permissible. There seems little doubt that fertilization could be completed in a normal manner in each of the two cells that have been formed, and then the chromosome complements would be appropriate for a normal 2-cell embryo. Subsequent development could well be normal right through to birth and beyond. The resulting individual would be a *chimaera* – a mixture of two different populations of cells arising from separate zygotic lines. The recent medical literature records several cases of chimaerism in people, but whether they originated in the way we have just described or by an alternative pathway is difficult to say. This involves the joining up (fusion) of two normal early embryos, an event that has often been brought about experimentally with mouse embryos. The fused embryos reorganize (regulate) so that duplicity disappears, and when returned to the uterus of a foster-mother develop as a unit. The single young mouse that is eventually born is of normal size, proportions and viability (see Book 2, Chapters 1 and 4).

PARTHENOGENESIS

The term parthenogenesis means literally virgin birth, and is applied to development that begins without the participation of a spermatozoon (as distinct from gynogenesis where activation is induced by a spermatozoon that takes no further part in the development). Parthenogenesis is a regular method of reproduction in some insects (for instance in the honey bee in which the drones are produced parthenogenetically and the workers and queens by fertilization) or as a seasonal phenomenon, as in the gall wasps. It can also be induced experimentally in some animals, as in the famous experiment of pricking a frog's egg with a needle, done by Bataillon early in this century. Since then there has been a good deal of research on the induction of parthenogenesis in mammals, and a degree of development occasionally going as far as half way through pregnancy has been obtained in mice and rabbits. There are a few reports of the birth of young after experimental induction, but none of these claims stands up to close inspection. Theoretically, the odds are heavily against the likelihood of parthenogenesis going to birth and beyond in mammals: the mammalian genome contains many thousands of genes that are each capable of mutating to a lethal allele. In parthenogenesis lethals would not be compensated for by normal genes, as they commonly are in fertilization. Accordingly the mammalian parthenogenone, the developing parthenogenetic embryo, would have very little chance of surviving to birth. However, the vast size of the human population compels the admission that with a one-in-a-million chance there could be a sprinkling of individuals in the community who have arisen by this means. They would of course be female, and would resemble their mothers very closely, but otherwise need not show any tell-tale features.

FERTILIZATION *IN VITRO*

The ease with which the eggs of several marine invertebrates and of some amphibians can be fertilized in the laboratory, by simply placing them with spermatozoa, has led people to suppose that the same trick can be done with mammalian material. Attempts were first made around 1880, and continued sporadically through the years, the reports usually being accompanied by the claim, expressed or implied, that fertilization had been obtained. In the light of contemporary knowledge, however, we can say with a good deal of confidence that success was not in fact achieved until about 1954. Only then were the implications of the need for spermatozoa to undergo capacitation properly appreciated, and experimenters took the precaution of utilizing spermatozoa recovered from the female tract some hours after mating, and adding these to the eggs held in a suitable medium *in vitro*.

The first successful experiments were made with rabbit germ cells by Charles Thibault and Louis Dauzier at Jouy-en-Josas, and a few years later M. C. Chang in Shrewsbury showed that the embryos obtained through fertilization *in vitro* were normal, by transferring them to the genital tracts of foster-mothers and observing the subsequent birth of healthy young rabbits (Fig. 5-12). Since then hamster and mouse gametes have been found

Fig. 5-12. Some of the first rabbits born from eggs fertilized *in vitro* and transferred to the oviduct of a foster-moster (in the experiments of M. C. Chang). The difference in coat colour helps to make it clear that the young rabbits are not derived from the foster-mother's eggs. (Modified from C. R. Austin, Fig. 17, Ch. VI in *Mechanisms Concerned with Conception*, Ed. C. G. Hartman. London; Pergamon (1963).)

more convenient. Both hamster and mouse spermatozoa readily undergo capacitation *in vitro*, provided the culture conditions are appropriate, while rabbit spermatozoa must reside for some hours in the female genital tract. Appropriate culture conditions can be achieved quite simply in a physiologically balanced salt solution to which a compound of high molecular weight, such as albumin or polyvinylpyrrolidone, is added. Human spermatozoa evidently resemble those of mouse and hamster, and not those of the rabbit, for they can complete their capacitation in such a medium *in vitro*. Once the preparation of the spermatozoa is thus completed, it is a comparatively simple matter to persuade them to penetrate and fertilize eggs *in vitro*, though again of course the precise conditions of the experiment are vitally important.

Quite recently the cat has been added to the list of animals whose eggs can be fertilized *in vitro*; the investigators used spermatozoa from the female tract but that this was really necessary remains to be shown. Presumably in due course the procedure can be successfully applied to the eggs of all other mammals, but species differences are remarkable and for some reason animals such as the rat have persistently refused to join the group of those with co-operative germ cells.

Embryos arising from eggs fertilized *in vitro* are normal, provided the eggs have completed their maturation in the ovary first. This has been clearly demonstrated in the rabbit and mouse by transferring such embryos to suitable host animals, and subsequently finding normal young born or normal full-term fetuses in the uterus. Attempts to obtain maturation *in vitro*, and then to follow this with fertilization *in vitro*, has led so far to disappointingly few normal embryos – we still need to know more about the necessary conditions for maturation *in vitro*. Human eggs recovered after maturation in the ovary and then fertilized *in vitro* seem able to develop normally in culture to the stage of blastocyst. At this point further progress requires transfer to a human recipient. There is no doubt that this step will be attempted shortly, and if successful the manoeuvre will complete an invaluable method for overcoming sterility due to blockage of the oviduct – a not very uncommon problem in young women today. The refinement and future potentialities of these techniques in man are examined in some detail in Book 5.

Fertilization plays rather a secondary role in reproduction, in the sense that this implies multiplication. With the fusion of egg and spermatozoon fertilization actually involves a reduction in number, and in many animals below mammals development can proceed normally in its absence. The primary role of fertilization lies in its genetic function – the bringing together or recombination of hereditary factors from two different individuals, and their reassortment in the genome of a new individual. Fertilization thus not only provides the mechanism for biparental inheritance but also prevents the progressive establishment of sub-groups, based on mutations, and so makes for integration of the race.

SUGGESTED FURTHER READING

Fertilization. C. R. Austin. Englewood Cliffs, N.J.; Prentice-Hall (1965).

Fertilization. 2 vols. Ed. C. B. Metz and A. Monroy. New York and London; Academic Press (1967, 1969).

Ultrastructure of Fertilization. C. R. Austin. New York; Holt Rinehart and Winston (1968).

La Fecondation chez les Mammifères. C. Thibault. In *Traité de Zoologie*, vol. XVL, pt. vi. Ed. P. P. Grassé. Paris; Masson et Cie (1969).

Schering Symposium on Mechanisms Involved in Conception, Berlin 1969. London and Braunschweig; Pergamon-Vieweg (1970).

Schering Symposium on Intrinsic and Extrinsic Factors in Early Mammalian Development, Venice, 1970. London and Braunschweig; Pergamon-Vieweg (1971).

Index

Index